T0281157

FASCINATING PROBLEMS FOR YOUNG PHYSICISTS

Problem-solving is the cornerstone of all walks of scientific research. *Fascinating Problems for Young Physicists* attempts to clear the boundaries of seemingly abstract physical laws and their tangible effects through a step-by-step approach to physics in the world around us. It consists of 42 problems with detailed solutions, each describing a specific, interesting physical phenomenon. Each problem is further divided into questions designed to guide the reader through, encouraging engagement with and learning the physics behind the phenomenon. By solving the problems, the reader will be able to discover, for example, what the relation is between the mass of an animal and its expected lifetime, or what the efficiency limit is of wind turbines. Intended for first-year undergraduate students and interested high school students, this book develops inquiry-based scientific practice and enables students to acquire the necessary skills for applying the laws of physics to realistic situations.

NENAD VUKMIROVIĆ has been a professor at the Institute of Physics Belgrade, Serbia, since 2010. His research interests are in theory and simulation of electronic properties of semiconductor materials and nanostructures. He is also very active in work with talented high school students. Nenad has been the author of many problems at high school physics competitions since 2012, including the problems at the final stage of competitions such as the Serbian Physics Olympiad.

VLADIMIR VELJIĆ received his PhD in physics in 2019 from the University of Belgrade and recently made a career switch from scientist to data scientist. Considering that his big passion is physics education and working with talented high school students, Vladimir has actively participated in the organization of several high school physics competitions on the national and international levels. Vladimir has been the author of theoretical problems at all levels of national competition.

FASCINATING PROBLEMS FOR YOUNG PHYSICISTS

Discovering Everyday Physics Phenomena and Solving Them

NENAD VUKMIROVIĆ

Institute of Physics Belgrade

VLADIMIR VELJIĆ

Institute of Physics Belgrade

CAMBRIDGE
UNIVERSITY PRESS

CAMBRIDGE
UNIVERSITY PRESS

University Printing House, Cambridge CB2 8BS, United Kingdom

One Liberty Plaza, 20th Floor, New York, NY 10006, USA

477 Williamstown Road, Port Melbourne, VIC 3207, Australia

314–321, 3rd Floor, Plot 3, Splendor Forum, Jasola District Centre, New Delhi – 110025, India

103 Penang Road, #05–06/07, Visioncrest Commercial, Singapore 238467

Cambridge University Press is part of the University of Cambridge.

It furthers the University's mission by disseminating knowledge in the pursuit of
education, learning, and research at the highest international levels of excellence.

www.cambridge.org
Information on this title: www.cambridge.org/9781009160285
DOI: 10.1017/9781009160261

First published 2022

A catalogue record for this publication is available from the British Library.

ISBN 978-1-009-16028-5 Hardback
ISBN 978-1-009-16027-8 Paperback

Additional resources for this publication at www.cambridge.org/fpyp.

Contents

Preface

Solving of problems has been an indispensable part in training of generations of physicists. A student deepens the understanding of the laws of physics when challenged to apply them to specific physical systems. There is a significant number of model systems where it is relatively straightforward to apply physical laws (a point particle in the constant field, a body on an inclined plane, a body connected to a spring, simple pendulum, ideal gases, system of point charges at rest, to name a few). The problems related to these and similar systems are part of every general physics course and can be found in general physics textbooks or in separate problem books. However, these systems remain relatively abstract and students do not necessarily get a good grasp of the fact that physics describes almost all phenomena in everyday life.

The aim of our book is to bridge this gap. Namely, we have selected the problems where the knowledge of general physics can be applied to phenomena from the world around us. By solving the problems in this book, the student will be able to discover, for example, why are the world records in high jump, long jump or pole vault at the specific lengths or heights, what is the relation between the mass of the animal and its expected lifetime, what is the efficiency limit of wind turbines, how do we see, why we see nice butterfly colors, etc. The book consists of 42 problems with detailed solutions, each describing a certain interesting physical phenomenon from everyday life. Each problem is divided into several (typically 5–15) questions that guide the reader throughout the problem and help the reader to understand the physics behind the phenomenon. Most of the problems in the book were prepared by the authors, while references are given in the case of few problems that were adapted from other sources.

The inspiration for the problems came from the world around us rather than from the branches of physics. For this reason, the problems are grouped by the systems from nature that they treat. First group of problems deals with the physics of the human body. The operation of all machines that we use or see in everyday life is based on physical laws and we address some of these machines in the next group of problems. The following group of problems addresses the physics behind various popular sport disciplines. Physical effects in the living world and in the natural phenomena such as rain, rainbow, lightning, earthquakes, and ocean waves are treated next. We then address the effects related to conversion of energy from one form to another and close the book with a group of several unrelated interesting problems.

Our intention in the problems was certainly not to give an exhaustive treatment of the phenomenon neither to exhaustively treat the phenomena from nature. The questions guide the reader to recognize and understand the basics of the main physical effects behind the phenomenon and we hope that these might stimulate the reader to investigate the phenomenon further. In this regard, the reader may find useful the Jupyter notebooks that accompany the book that contain all calculations necessary to obtain the final numerical result. The reader can use these, for example, to change the parameters of the system and see what would be the effect of these changes.

Intended audience for this book are talented high school students who are motivated to deepen their knowledge of physics and first and second year undergraduate students that take courses of general physics. The book will certainly be of use to high school teachers that work with talented students, while some easier parts of the problems can be also used as examples in regular high school classes. At the university level, the problems from the book can be used in lectures as nice examples of applications of the laws of physics in life and as homework assignments to students.

Physical laws and mathematical techniques that should be applied to solve the problems are part of high school curriculum in most countries and are in-line with the curriculum of International Physics Olympiad for high school students. The difficulty of the questions varies. Some introductory questions in the problem are easy and can be solved using direct application of formulas that describe certain physical laws, while most of the problems contain parts that are quite challenging and are of difficulty comparable to most difficult problems at International Physics Olympiad for high school students. In fact, a few of these problems were posed by authors at national physics competitions in Serbia. We would like to acknowledge our colleagues who have acted as reviewers for these problems or with whom we have discussed the problems: Veljko Janković, Dimitrije Stepanenko, Antun Balaž, Božidar Nikolić, Duško Latas, Ana Hudomal, Darko Tanasković, Dejan Đokić, Marko Opačić, Mihailo Rabasović, Aleksandar Krmpot, Nikola Jovančević, Petar Mali, Vladan Pavlović, Marko Kuzmanović, and Milan Radonjić.

Finally, we would like to give a few suggestions for students on how to use this book. The problems can be solved in any order, while the problems within a certain group were sorted from somewhat easier to more difficult. We strongly encourage the students to make a significant effort first to solve the problem without consulting the solution. Learning in the process of trying to solve the problem is indispensable, even when the student eventually does consult the solution. When consulting the solution, we suggest to use the solution first only as a hint and go back to solve the problem using this hint.

We hope that the problems from this book will guide the students to recognize the laws of physics in the world around us, that they will deepen their knowledge of physics by solving the problems and that this book will introduce a fresh perspective to the use of problem solving in physics teaching.

1

Human

Problem 1   Human Eye

How do we see?
What kind of glasses might we need?
When can we distinguish between the two eyes of a cat during the night?

A schematic view of the structure of the human eye is presented in Figure 1.1. Light rays that refract at the cornea and eye lens end up at the retina, which produces nerve impulses sent to the brain down the optic nerve. In a simplified model of an eye, the cornea and eye lens can be replaced with one converging lens (called simply the lens in the remainder of the text) while the retina can be modeled as a disk of radius $R = 1.00\,\text{cm}$, the axis of which coincides with the optical axis of

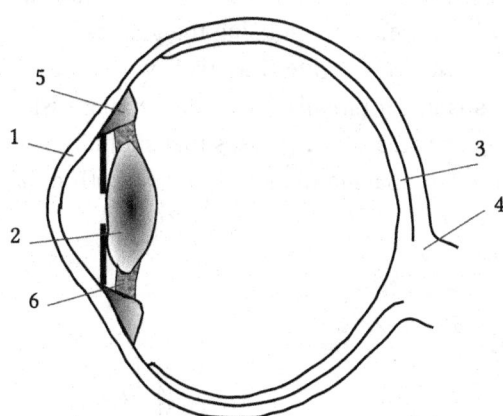

Figure 1.1 Scheme of the structure of the human eye: (1) cornea, (2) eye lens, (3) retina, (4) optic nerve, (5) ciliary muscles, (6) suspensory ligament

the lens, as shown in Figure 1.2. The distance between the retina and the lens is $d = 2.40\,\text{cm}$. A human can adjust the focal length of the lens and therefore has the capability of clearly seeing objects at different distances. This process is called eye accommodation and is enabled by ciliary muscles connected to the eye lens by a suspensory ligament. These muscles act to tighten or relax the ligaments and therefore thin down or thicken the lens. Consequently the focal length of the lens changes.

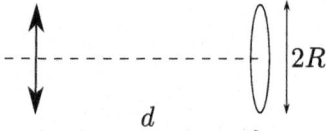

Figure 1.2 A simplified model of the human eye

(a) A human has regular eyesight if images of all objects from a distance larger than $d_0 = 25.0\,\text{cm}$ can be formed at the retina. What is the range of the lens' focal lengths for a human with regular eyesight?

(b) The maximal focal length $f_{\max}$ of the lens for a nearsighted man is smaller than the upper limit of the range determined in part (a). This man uses glasses with a diopter value of $D_1 = -1.00\,\text{m}^{-1}$ to clearly see very distant objects. Determine $f_{\max}$ and find the maximal distance of an object that this man can clearly see without using the glasses. For simplicity neglect the distance between the glasses and the lenses.

(c) The minimal focal length $f_{\min}$ of the lens for a farsighted woman is larger than the lower limit of the range determined in part (a). This woman needs glasses with a diopter value of $D_2 = 2.00\,\text{m}^{-1}$ to clearly see objects at a distance of $d_0 = 25.0\,\text{cm}$. Determine $f_{\min}$ and find the minimal distance of an object that this woman can clearly see without using the glasses.

(d) A person is nearsighted (farsighted) as well when the distance between the retina and the lens is larger (smaller) than the regular distance of $d = 2.40\,\text{cm}$. Calculate the diopter value of the glasses that should be used by a man with a distance between the retina and the lens of $d_1 = 2.50\,\text{cm}$ ($d_2 = 2.30\,\text{cm}$).

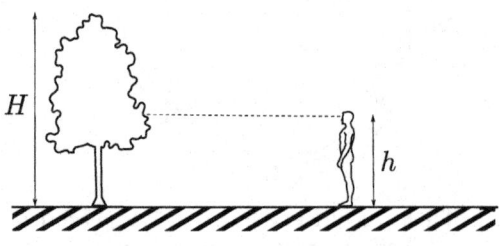

Figure 1.3 With problem 1(e)

(e) A man with regular eyesight whose height is $h = 2.00$ m is observing a tree of height $H = 2h$ (Figure 1.3). His view is directed toward the middle of the tree. What is the minimal distance between the man and the tree that allows him to see the whole tree?

Two types of light receptors are placed at the retina – rods (about $N_1 = 10^8$ of them) and cones (about $N_2 = 6 \cdot 10^6$ of them). Rods enable night vision, while cones are used for vision during the day. Assume that a person can distinguish two distant objects during the day (night) if their images are at different cones (rods). Assume also that the cones (rods) are evenly distributed on the retina surface and that their positions form a square lattice.

(f) Two point objects are at a mutual distance of $a = 1.00$ mm. The direction that connects them is perpendicular to the optical axis of the lens (Figure 1.4). What is the maximal distance from which a woman can distinguish between these two objects during the day?

Figure 1.4 With problem 1(f)

(g) At what maximal distance can a woman read the license plates of a car during the day? Assume that the license plates can be read if a woman can distinguish between the point objects at a mutual distance of $a = 1.00$ cm.

(h) At what maximal distance can a woman distinguish between the two eyes of a cat during the night? The eyes of a cat are at a mutual distance of $a = 2.00$ cm.

Solution of Problem 1

(a) To see an object at a distance p from the eye, a human needs to accommodate the focal length of the lens so that the image of the object is formed at the retina (which is at a distance $l = d$ from the lens). For an object at a distance $p_1 = d_0$ the focal length is given by lens equation $\frac{1}{f_1} = \frac{1}{p_1} + \frac{1}{l}$. For an object at a distance $p_2 \to \infty$ we have $\frac{1}{f_2} = \frac{1}{p_2} + \frac{1}{l}$. From previous equations we obtain $f_1 = 2.19$ cm and $f_2 = 2.40$ cm. Consequently the lens focal length of a human with regular eyesight takes a range from $f_1 = 2.19$ cm to $f_2 = 2.40$ cm.

(b) The lens focal length and the distance of the object that the man clearly sees are related by $\frac{1}{f} = \frac{1}{p} + \frac{1}{d}$. Consequently, without the use of glasses, this man cannot clearly see objects at a distance larger than p_{max}, where

$$\frac{1}{f_{max}} = \frac{1}{p_{max}} + \frac{1}{d}. \tag{1.1}$$

The focal length of the system lenses-glasses f_{ns} satisfies the relation $\frac{1}{f_{ns}} = \frac{1}{f} + D_1$. When this man clearly sees very distant objects with the use of glasses, the lens equation reads

$$\frac{1}{f_{max}} + D_1 = \frac{1}{p_2} + \frac{1}{d}, \tag{1.2}$$

where $p_2 \to \infty$. From equation (1.2) we obtain $f_{max} = \frac{d}{1-dD_1} = 2.34$ cm. By subtracting equations (1.1) and (1.2) we find $p_{max} = -\frac{1}{D_1} = 1.00$ m.

(c) Without the use of glasses, this woman cannot clearly see objects at a distance smaller than p_{min}, where

$$\frac{1}{f_{min}} = \frac{1}{p_{min}} + \frac{1}{d}. \tag{1.3}$$

The lens equation for a woman with glasses looking at an object at a distance d_0 reads

$$\frac{1}{f_{min}} + D_2 = \frac{1}{d_0} + \frac{1}{d}. \tag{1.4}$$

From equation (1.4) it follows that

$$f_{min} = \frac{1}{\frac{1}{d_0} + \frac{1}{d} - D_2} = 2.29 \text{ cm}. \tag{1.5}$$

By subtracting equations (1.3) and (1.4) we obtain

$$p_{min} = \frac{d_0}{1 - D_2 d_0} = 50.0 \text{ cm}. \tag{1.6}$$

(d) The lens equation for a man with regular distance between the lens and the retina when he clearly sees an object at a distance p is $\frac{1}{f} = \frac{1}{p} + \frac{1}{d}$. For a man with distance d_i between the retina and the lens who uses glasses with diopter value D_i and clearly sees the same object when the lens focal length is the same, we obtain $\frac{1}{f} + D_i = \frac{1}{p} + \frac{1}{d_i}$. Subtracting the previous two equations, we find $D_i = \frac{1}{d_i} - \frac{1}{d}$. Consequently, we find in the first case $D_1 = -1.67 \text{ m}^{-1}$ and in the second case $D_2 = 1.81 \text{ m}^{-1}$.

(e) A man sees the whole tree when the size L of the image of the tree on the retina is smaller than the retina diameter (Figure 1.5). Using the similarity of the triangles in Figure 1.5 we obtain $\frac{L}{H} = \frac{d}{x}$, where x is the distance between the man and the tree. Consequently the man sees the whole tree when $L = H\frac{d}{x} < 2R$, leading to $x > \frac{Hd}{2R} = 4.80$ m.

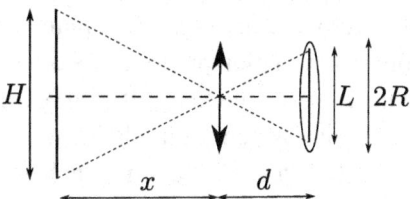

Figure 1.5 With the solution of problem 1(e)

(f) The number of cones per unit surface is equal to $N_S = \frac{N_2}{R^2\pi}$. On the other hand, since we assume that the positions of cones form a square lattice with lattice constant b, we also have $N_S = \frac{1}{b^2}$. From the previous two equations it follows that $b = R\sqrt{\frac{\pi}{N_2}} = 7.24\,\mu$m. When the woman is at a maximal distance at which she can still distinguish between the two objects, the images of the objects are formed at two neighboring cones. From the similarity of triangles in Figure 1.6, we find $\frac{a}{x} = \frac{b}{d}$ – that is, $x = \frac{ad}{b} = 3.32$ m.

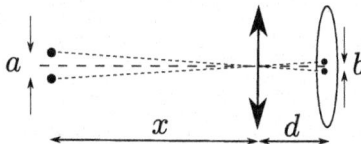

Figure 1.6 With the solution of problem 1(f)

(g) From the solution of part (f) we have $x = \frac{ad}{b}$, where in this case $a = 1.00$ cm, leading to $x = 33.2$ m.

(h) Since the woman observes the cat during the night, the solution of part (f) is modified only by replacing the number of cones with the number of rods. Consequently, $x = \frac{ad\sqrt{N_1}}{R\sqrt{\pi}} = 271$ m.

We refer the reader interested in more details regarding the physics of the human eye to chapter 12, reference [13].

Problem 2   The Circulation of Blood

How powerful is the human heart?
How does a bypass help in the case of arteriosclerosis?

The human cardiovascular system consists of the heart, the blood, and the blood vessels. The heart pumps the blood through the blood vessels. The blood carries nutrients and oxygen to and carbon dioxide away from various organs. The most important portions of the cardiovascular system are pulmonary circulation and systemic circulation. Pulmonary circulation pumps away oxygen-depleted blood from the heart via the pulmonary artery to the lungs. It then returns oxygenated blood to the heart via the pulmonary vein. Systemic circulation transports oxygenated blood away from the heart through the aorta. The aorta branches to arteries that bring the blood to the head, the body, and the extremities. The veins then return oxygen-depleted blood to the heart. The direction of blood flow is determined by four heart valves. Two of them are positioned between the antechambers and the chambers, while two are located between the chambers and the arteries.

(a) The heart pumps blood by contraction of the muscles of the antechambers and chambers. The blood pressure gradually increases from the minimal (diastolic) value of $p_d = 80\,\mathrm{mmHg}$ to the maximal (systolic) value of $p_s = 120\,\mathrm{mmHg}$ during contraction.

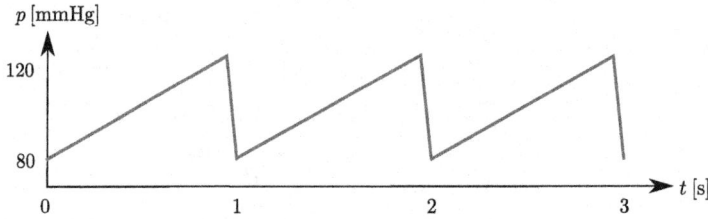

Figure 1.7 The graph of the dependence $p(t)$

The muscle then relaxes and the value of pressure suddenly decreases, as shown in Figure 1.7. The heart contracts (beats) around 60 times a minute. Each contraction pumps around 75 ml of blood. The pump shown in Figure 1.8 is a simple model of the heart. The heart decreases the volume during the contraction, which corresponds to the upward motion of the piston in the model.

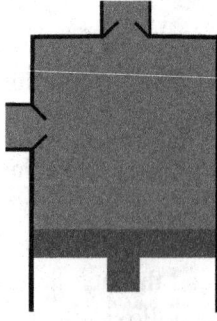

Figure 1.8 A pump as a model of the heart

Thereby the pressure increases and closes the input valves while it opens the output valves. Determine the power of the heart.

The boundary between laminar and turbulent flow of blood is determined from the Reynolds number, which is directly proportional to the speed of blood v. The Reynolds number is a dimensionless quantity that depends as well on the density of blood $\rho = 1,060\,\text{kg/m}^3$, viscosity of blood $\eta = 4.0 \cdot 10^{-3}\,\text{Pa} \cdot \text{s}$ and the diameter of the blood vessel D. The flow is turbulent if the Reynolds number is larger than 2,000, while it is laminar otherwise.

(b) Derive the expression for the Reynolds number using dimensional analysis. Assume that the dimensionless constant that appears in front of the expression is equal to 1.

(c) The diameter of the aorta is $D = 10\,\text{mm}$. Calculate the maximal speed of laminar blood flow in the aorta.

We consider next the laminar flow of blood through the artery whose shape is a cylinder of length L and radius R, as shown in Figure 1.9. The flow of blood in the artery is caused by the difference of pressures Δp at the ends of the artery, which is a consequence of blood pumping from the heart. The blood does not slide at the walls of the artery. For this reason, a cylindrical layer of blood that is at rest is formed near the wall of the artery. The viscosity of the blood causes laminar flow where each layer slides between neighboring layers. The viscosity force between the layers F is given by Newton's law,

$$F = \eta S \frac{\Delta v}{\Delta r},$$

where η is the viscosity of the blood, S is the area of the layer that is in contact with the neighboring layer, and $\Delta v/\Delta r$ is the gradient of speed in the radial direction. The walls of the artery are inelastic and the speed of flow does not change between the points on the same line in the direction of the artery.

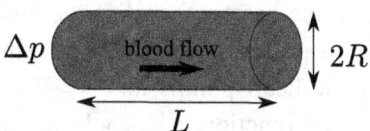

Figure 1.9 Artery

(d) Determine the dependence of the speed of blood on the distance from the artery axis.

(e) Using the analogy of electrical resistance, one can define the resistance of blood flow as the ratio of the pressure difference and the volume flow caused by this difference of pressures. Determine the blood flow resistance through the artery.

(f) As a consequence of arteriosclerosis, the inner diameter of a part of the artery decreased from $d_1 = 6.0$ mm to $d_2 = 4.0$ mm. How many times was the blood flow resistance increased in this part of the artery? To reduce the blood flow resistance, a bypass can be introduced. A healthy artery or vein is removed from another part of the patient's body and attached in parallel to this part of the artery. Assume that the bypass is of the same length as this part of the artery. How many times does the blood flow resistance decrease after the introduction of a bypass of diameter $d_3 = 5.0$ mm?

When the blood enters the artery, the speed of the blood is nearly the same throughout the cross-section of the artery. This means that the blood needs to accelerate and decelerate to reach the regime considered in previous parts of the problem. The blood near the artery walls decelerates to zero speed, while the part in the center of the artery accelerates to the maximal value of the speed. Consider the situation when we neglect the viscosity and when the blood accelerates along the artery.

(g) Determine the relation between the pressure difference Δp at the ends of the artery and the change of volume flow $\Delta q / \Delta t$ as a function of blood density ρ, the length of the artery L, and its radius R.

(h) As in part (e), the analogy with electrical circuits can be also introduced in part (g). Which element of the electric circuit can be used to describe the relation determined in part (g)?

Solution of Problem 2

(a) The work performed by the pump when the piston moves by Δr is

$$\Delta A = F \Delta r = \frac{F}{S} \Delta r S = p \Delta V . \tag{1.7}$$

The work performed by the heart is equal to the area under the graph of the function $p(V)$. The heart performs 60 beats per minute, which is 1 beat per second. Consequently the heart pumps in $V = 75$ ml of blood each second. Therefore, the graph of the function $p(V)$ looks as shown in Figure 1.10. The work performed by the heart during 1 beat is

$$A = p_d V + \frac{1}{2}(p_s - p_d)V = \frac{1}{2}(p_s + p_d)V = 1.0\,\text{J} . \tag{1.8}$$

The work A is performed by the heart during $t = 1$ s, which means that the corresponding power is $P = A/t = 1\,\text{J}/1\,\text{s} = 1.0\,\text{W}$.

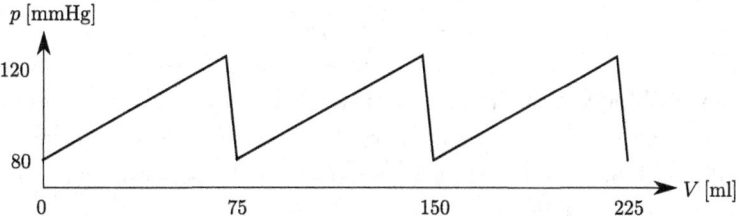

Figure 1.10 The graph of the function $p(V)$

(b) We can find the expression for the Reynolds number using dimensional analysis

$$\text{Re} = v\rho^\alpha \eta^\beta D^\gamma \Rightarrow 1 = [\text{m} \cdot \text{s}^{-1}] [\text{kg} \cdot \text{m}^{-3}]^\alpha [\text{kg} \cdot \text{m}^{-1}\text{s}^{-1}]^\beta [\text{m}]^\gamma, \qquad (1.9)$$

which leads to the system of equations

$$1 - 3\alpha - \beta + \gamma = 0, \, -1 - \beta = 0, \, \alpha + \beta = 0, \qquad (1.10)$$

whose solution is $(\alpha, \beta, \gamma) = (1, -1, 1)$. Therefore, the Reynolds number is given by the expression

$$\text{Re} = \frac{\rho v D}{\eta}. \qquad (1.11)$$

(c) The maximal speed of blood in the aorta is obtained for $\text{Re} = 2,000$ and reads

$$v = \frac{\eta \text{Re}}{\rho D} = 75 \, \frac{\text{cm}}{\text{s}}. \qquad (1.12)$$

The Reynolds number reaches the critical value when the valves of the aorta open. The blood is then under big pressure and reaches a speed as high as $120 \, \text{cm/s}$. So-called Korotkoff sounds appear then as a consequence of turbulent flow. These can be heard using a stethoscope. This fact is used when blood pressure is measured using a sphygmomanometer.

(d) The system has cylindrical symmetry. Consequently the speed of blood is constant in each thin cylindrical layer. Consider the part of blood in the shape of a cylinder of radius r. This part of blood in the artery moves due to pressure difference Δp, which yields the force $F_1 = \pi r^2 \Delta p$. The magnitude of the viscosity force that acts on this layer is $F_2 = 2\pi r L \eta \frac{\Delta v}{\Delta r}$. Since each layer of blood is moving at a constant velocity, we obtain from Newton's first law that

$$F_1 = F_2 \Rightarrow \Delta v = \frac{\Delta p}{2\eta L} r \Delta r. \qquad (1.13)$$

By transforming the equation (1.13) to differential form and performing integration with the boundary condition $v(R) = 0$, we obtain the dependence of the speed of blood on the distance from the axis of the artery:

$$v(r) = \frac{\Delta p}{4\eta L}(R^2 - r^2). \tag{1.14}$$

(e) The flow of blood through the ring of width dr, which is located in the region between r and $r+dr$, is $v(r)dS$, where $dS = 2\pi r dr$ is the area of that ring. The flow of blood through the artery is then obtained by performing the integration over all rings, which leads to

$$q = \int_0^R v(r)2\pi r dr = \int_0^R \frac{\pi\Delta p}{2\eta L}(rR^2 - r^3)dr = \frac{\pi\Delta p R^4}{8\eta L}, \tag{1.15}$$

and consequently the blood flow resistance is

$$\mathcal{R} = \frac{\Delta p}{q} = \frac{8\eta L}{\pi R^4}. \tag{1.16}$$

(f) Due to arteriosclerosis the blood flow resistance in the sick part of the artery $\mathcal{R}_2$ increases in comparison to the resistance in the healthy artery $\mathcal{R}_1$, which leads to

$$\frac{\mathcal{R}_2}{\mathcal{R}_1} = \left(\frac{d_1}{d_2}\right)^4 = 5.1. \tag{1.17}$$

After the bypass is introduced, the sick part of the artery and the bypass form a parallel connection of two resistors with equivalent resistance $\mathcal{R}_e$. The resistance then reduces by

$$\frac{\mathcal{R}_2}{\mathcal{R}_e} = \frac{\mathcal{R}_2}{\frac{\mathcal{R}_2\mathcal{R}_3}{\mathcal{R}_2+\mathcal{R}_3}} = 1 + \left(\frac{d_3}{d_2}\right)^4 = 3.4. \tag{1.18}$$

(g) Newton's second law applied to the blood in the artery gives:

$$m\frac{\Delta v}{\Delta t} = \Delta p S, \tag{1.19}$$

where $m = \rho V = \rho L \pi R^2$ is the mass of the blood in the artery, $\Delta v/\Delta t$ is the change of the speed of blood along the artery, and $S = \pi R^2$ is the area of the inner cross-section of the artery. The change of flow is $\Delta q = \Delta(R^2\pi v) = \pi R^2 \Delta v$, which along with equation (1.19) gives

$$\Delta p = \left(\frac{\rho L}{\pi R^2}\right)\frac{\Delta q}{\Delta t}. \tag{1.20}$$

(h) One can conclude from part (e) that the change of pressure is analogous to the potential difference, while the flow of blood is analogous to the electrical current. Consequently equation (1.20) is analogous to the equation

$$\Delta\varphi = \mathcal{L}\frac{\Delta I}{\Delta t}, \tag{1.21}$$

which leads to the conclusion that the inductor is the analogous electrical component that describes the blood flow in part (g).

We refer the reader interested in more details about the circulation of blood in the human body to references [23] and [41].

Problem 3   A Human As a Heater

How many persons are needed to heat a room to the same temperature as a heater?

You will have certainly noticed that it can be very hot in a room where a lot of people are present. The reason for this is that humans emit heat and consequently they heat the room. We assume that humans exchange heat with the room only by thermal radiation and heat conduction.

We consider first how a human exchanges heat with their surroundings by thermal radiation. We use a simplified model in which the shape of the human is a ball of radius r, while the room is a sphere of radius R, which is much larger than r. The centers of the ball and the sphere coincide, as shown in Figure 1.11(b). Assume that the human and the internal walls of the room radiate as black bodies, whose temperatures are respectively T_c and T_z.

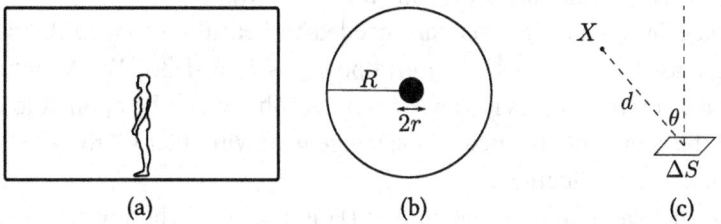

Figure 1.11 (a) A human in the room. (b) Simplified model of the human and the room. (c) A scheme accompanying equation (1.22)

To solve the problem you can make use of the following fact. In point X the intensity of electromagnetic radiation emitted by a black body in the shape of a small flat tile is given as

$$I = \frac{\sigma T^4 \Delta S \cos \theta}{d^2 \pi},$$ (1.22)

where T is the temperature of the body, ΔS is the area of one side of the tile, d is the distance between the tile and point X, and θ is the angle between the direction connecting point X with the tile and the direction perpendicular to the plane of the tile, as shown in Figure 1.11(c).

(a) Determine the expression for the power of the radiation that the human emits.

(b) Determine the expression for the power of the radiation that the human absorbs.

(c) Determine the expression for the power of the radiation that the human exchanges with their surroundings.

(d) Calculate the power of the radiation that the human exchanges with their surroundings. Assume that the equations derived in previous parts of the problem can be applied to a human in the room. Use the following numerical values in this part of the problem – $t_c = 36.0°C$, $t_z = 20.0°C$ – and assume that the surface area of one human is equal to $S_c = 1.90\,\text{m}^2$. The Stefan–Boltzmann constant is $\sigma = 5.67 \cdot 10^{-8}\,\text{Wm}^{-2}\text{K}^{-4}$, $0°C = 273.15\,\text{K}$.

The law of heat conduction states that the amount of heat that a human exchanges with their surroundings in unit time by heat conduction is given by the expression

$$\frac{dQ}{dt} = \alpha S_c (T_o - T_c), \tag{1.23}$$

where α is the coefficient of heat conduction equal to $\alpha = 4.50\,\frac{\text{W}}{\text{m}^2\cdot\text{K}}$ for a human wearing regular clothes, S_c is the surface area of the human, and T_o is the temperature of the surroundings.

(e) Calculate the power of heat that a human from part (d) exchanges with their surroundings by heat conduction. Assume that the temperature of the surroundings is equal to the temperature of walls from part (d).

(f) Calculate how many persons are needed to heat the room to the same temperature as the heater, whose useful power is $P_g = 1.30\,\text{kW}$. Assume that the formulas derived in previous parts of the problem can be applied to each person in the room and that people exchange heat with their surroundings only by radiation and conduction.

(g) Answer the same question as in part (f) in the case when people in the room wear winter clothes whose coefficient of heat conduction is $\alpha' = 2.40\,\frac{\text{W}}{\text{m}^2\cdot\text{K}}$.

Solution of Problem 3

(a) Since the human emits as a black body, the power of the emitted radiation is given by the Stefan–Boltzmann law $P_{em} = \sigma T_c^4 S_c$, where $S_c = 4r^2\pi$ is the surface area of the human, which leads to $P_{em} = \sigma T_c^4 4r^2\pi$.

(b) We determine first the power of the radiation emitted by a small part of the wall of area ΔS_z that is absorbed by the human. We denote this power as ΔP_{pr}. The intensity of the radiation emitted by that part of the wall at a distance R is, according to the formula given in the problem, $\Delta I = \frac{\sigma T_z^4 \Delta S_z}{R^2 \pi}$. The radiation incident on the human is the radiation that would be incident on a tile of area $r^2\pi$ that is perpendicular to the direction that connects the part of the wall and the human. Therefore, $\Delta P_{\text{pr}} = \Delta I \cdot r^2 \pi$, which leads to $\Delta P_{\text{pr}} = \frac{\sigma T_z^4 \Delta S_z r^2}{R^2}$. The total power of the radiation absorbed by the human is obtained by adding all ΔP_{pr}. Bearing in mind that $S_z = 4R^2\pi$, we obtain $P_{\text{pr}} = \sigma T_z^4 4\pi r^2$.

(c) The power that the human exchanges with their surroundings is $P_{\text{r}} = P_{\text{em}} - P_{\text{pr}} = \sigma \left(T_c^4 - T_z^4\right) 4\pi r^2$.

(d) Using the expression from part (c) leads to $P_{\text{r}} = \sigma \left(T_c^4 - T_z^4\right) S_c$, and consequently $P_{\text{r}} = 188\,\text{W}$.

(e) The power of heat that the human exchanges with their surroundings by heat conduction is $P_{\text{t}} = \alpha S_c (T_o - T_c) = 137\,\text{W}$.

(f) To heat the room to the same temperature as the heater, the power that people exchange with their surroundings should be the same as the power of the heater. Consequently, $P_{\text{g}} = N\left(P_{\text{r}} + P_{\text{t}}\right)$, where N is the number of persons in question. Using the solutions of parts (d) and (e), we find $N \approx 4$.

(g) The power of heat that a human exchanges with their surroundings by heat conduction is now $P_{\text{t}}' = \alpha' S_c (T_o - T_c) = 73\,\text{W}$, and therefore $N = \frac{P_{\text{g}}}{P_{\text{r}} + P_{\text{t}}'} \approx 5$.

We refer the interested reader to reference [16]. Part of this problem was given by the authors at the national physics competition for the fourth grade of high school in Serbia in 2016.

Problem 4   Human Walk

How fast can humans walk without running?
What is the energy required for walking?

We consider the human walk in this problem. You might have noticed that it is quite difficult to walk fast without actually running.

(a) Make an order of magnitude estimate of the highest possible speed of the human walk without using the data given in the rest of the problem.

In the rest of this problem we consider the following model of the human walk and use it for a more detailed analysis of the human walk. We assume that the human body consists of three rigid rods that represent the two legs and the abdomen. The pelvis is where the legs and abdomen merge. The position of the abdomen during the walk is always vertical. One leg (which we call the standing leg) is always in contact with the ground, while the foot of the other leg is slightly above the ground but not in contact with the ground. Static friction between the legs and the ground is enough to prevent the standing leg from slipping. Unless otherwise stated, assume that muscles do not perform work and that the person moves only under the influence of external forces.

Different positions that a person takes during one step are presented in Figure 1.12. At the beginning of the step (moment 1) the angle between each leg and the vertical is $\theta = \theta_0$ ($\theta_0 < 48°$). The pelvis rotates around the point of contact of the standing leg and the ground and consequently the angle θ decreases until the legs become parallel (moment 3). Next, the angle θ increases to the value $\theta = \theta_0$ (moment 5). At moment 5 there is a change of standing leg as follows. First, the standing leg loses contact with the ground (leg D in Figure 1.12 [moment 5]). Immediately after that the second leg (leg L in Figure 1.12 [moment 5′]) becomes the standing leg.

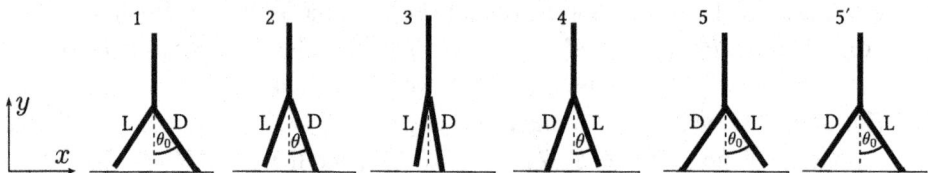

Figure 1.12 Positions a person takes during one step in five different moments of time (denoted as 1 to 5 and 5′). L denotes the left leg while D denotes the right leg

We assume for simplicity that the person's total mass is located in the pelvis. The length of each leg is l, the person's mass is m, and gravitational acceleration is g. The speed of the person at the moment when the legs are parallel is v_0.

(b) Find the expressions for the magnitude of angular velocity of the leg and for the reaction force of the ground at the moment when the angle between the leg and the vertical is θ.

(c) Determine the condition that the quantities v_0, g, l, and θ_0 should satisfy to make sure that the standing leg remains in contact with the ground.

(d) Calculate the maximal speed of walking for a person with short steps (small value of θ_0). The length of the leg is $l = 90$ cm while gravitational acceleration is $g = 9.81 \frac{m}{s^2}$. Compare the result with the average speed of the best athlete competing in a 10 km walking race. The world record in this discipline is 37 minutes and 11 seconds for male athletes and 41 minutes and 4 seconds for female athletes. Comment on the difference in the results.

(e) The angular velocity of the leg suddenly changes during the change of the standing leg. Find the ratio of the angular velocity of the leg immediately after and immediately before the change of standing leg.

(f) Determine the expression for the loss of kinetic energy of the person during the change of the standing leg.

(g) Calculate the work per distance traveled that muscles should perform in order to compensate for the loss of kinetic energy described in part (f). Use the following numerical values: $\theta_0 = 10°$, $v_0 = 2.0\,\frac{m}{s}$, and $m = 75\,kg$, while g and l are given in part (d).

(h) How much spinach should the person eat to have enough energy to walk the distance of $s = 10\,km$? The energy value of $100\,g$ of spinach is $28\,kcal$, where $1\,kcal = 4.196\,kJ$. Use the numerical values from parts (d) and (g).

Solution of Problem 4

(a) Assume that the person walks at a speed v. For an order of magnitude estimate, we assume that the person's whole mass m is located in the pelvis. During one step the pelvis performs the motion on a circle of radius l equal to the length of the leg. The person stays in contact with the ground if the force of gravity is larger than the centripetal force $mg > \frac{mv^2}{l}$, where g is gravitational acceleration. We therefore obtain $v > \sqrt{gl}$. Assuming that $g \approx 10\,\frac{m}{s^2}$ and that the length of the leg is $l \approx 1\,m$, we obtain a maximal speed of human walking of $v_{max} = \sqrt{gl} \sim 3\,\frac{m}{s}$.

(b) The trajectory of the pelvis is the circle of radius l, which implies that its speed is $v = l\omega$, where ω is the magnitude of the angular velocity of the leg. Since we assume in this problem that the person's whole mass is located in the pelvis, the person's kinetic energy is $T(\theta) = \frac{1}{2}mv(\theta)^2 = \frac{1}{2}ml^2\omega(\theta)^2$. The gravitational potential energy (with the reference level set at the ground) is $U(\theta) = mgl\cos\theta$. The law of energy conservation reads $U(\theta) + T(\theta) = U(0) + T(0)$ and consequently

$$\omega(\theta)^2 = \frac{v_0^2}{l^2} + \frac{2g}{l}(1 - \cos\theta). \tag{1.24}$$

Newton's second law of motion of the person in the y direction gives

$$\frac{dp_y}{dt} = N - mg, \tag{1.25}$$

where N is the magnitude of the reaction force of the ground and p_y is the y component of the person's momentum. The y coordinate of the person is given as $y = l\cos\theta$, which leads to $p_y = m\frac{dy}{dt} = -ml\dot{\theta}\sin\theta$, where we introduce the notation $\dot{\theta} = \frac{d\theta}{dt}$. Further differentiation leads to

$$\frac{dp_y}{dt} = -ml\left(\dot{\theta}^2\cos\theta + \sin\theta\frac{d\dot{\theta}}{dt}\right). \tag{1.26}$$

We further obtain

$$\frac{d\dot{\theta}}{dt} = \frac{d\dot{\theta}}{d\theta}\frac{d\theta}{dt} = \dot{\theta}\frac{d\dot{\theta}}{d\theta} = \frac{1}{2}\frac{d\dot{\theta}^2}{d\theta}. \tag{1.27}$$

Since the relation $\omega(\theta)^2 = \dot{\theta}^2$ holds at every moment of time, by using the trigonometric identity $\sin^2\theta = 1 - \cos^2\theta$, we obtain from equations (1.24) to (1.27) that

$$N = 3mg\cos^2\theta - 2mg\cos\theta - m\frac{v_0^2}{l}\cos\theta. \tag{1.28}$$

(c) The standing leg will remain in contact with the ground if the condition $N > 0$ is satisfied at every moment of time. This leads to the condition $v_0^2 < gl(3\cos\theta - 2)$. This condition is satisfied for each value of θ during the motion if the condition $v_0^2 < gl(3\cos\theta_0 - 2)$ is satisfied.

(d) Since a walk with short steps implies $\theta_0 \approx 0$, the solution of part (c) yields $v_0^{max} = \sqrt{gl} = 3.0\frac{m}{s}$. The average speed of world record holders in walking is $v_{sr} = \frac{s}{t}$, where $s = 10\,km$ and $t = 2,231\,s$ for males, while $t = 24,64\,s$ for females. Consequently $v_{sr} = 4.5\frac{m}{s}$ for males and $v_{sr} = 4.1\frac{m}{s}$ for females. This result implies that the average speed of world record holders is faster than this model predicts. One reason for that could be the simplicity of the model used, which gives only an estimate of maximal possible walking speed. Another reason is that athletes perform characteristic moves of the pelvis that enable walking at higher speeds while keeping the standing leg on the ground. This is particularly important in walking events because the athlete gets a warning every time the leg loses contact with the ground and three such warnings lead to disqualification.

(e) During the change of the standing leg (from moments 5 to 5′) it loses contact with the ground first, which does not lead to a change in the person's speed. When the second leg makes contact with the ground, a sudden change takes place in the person's speed and in the angular velocity of the leg. The person's angular momentum with respect to the point of contact of the standing leg with the ground (point O in Figure 1.13) is conserved during such contact. The reason for this is that gravity is the only external force acting during the contact that has torque with respect to point O. Since the duration of the contact is short and the force of gravity is finite, the change of angular momentum with respect to point O is negligible.

Angular momentum with respect to point O before the contact of the leg with the ground is $\vec{L}_i = m\vec{r} \times \vec{v}_i$, where $\vec{r} = -l\sin\theta_0\vec{e}_x + l\cos\theta_0\vec{e}_y$ is the position of the pelvis with respect to point O; $\vec{v}_i = l\omega_i\cos\theta_0\vec{e}_x - l\omega_i\sin\theta_0\vec{e}_y$, where $\vec{e}_x$ and $\vec{e}_y$ are the unit vectors in the x and y directions; $\vec{v}_i$ is the velocity of the pelvis just before the contact of the leg with the ground (Figure 1.13);

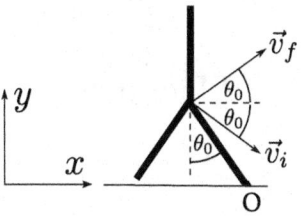

Figure 1.13 With solution of problem 4(e)

and ω_i is the magnitude of the angular velocity of the leg just before its contact with the ground. Using the identity $\cos(2\theta_0) = \cos^2\theta_0 - \sin^2\theta_0$, it follows that $(L_i)_z = -ml^2\omega_i\cos(2\theta_0)$. Next we have $\vec{L}_f = m\vec{r} \times \vec{v}_f$, where $\vec{v}_f = l\omega_f\cos\theta_0\vec{e}_x + l\omega_f\sin\theta_0\vec{e}_y$, with $\vec{v}_f$ being the velocity of the pelvis just after the contact of the leg with the ground (Figure 1.13), and ω_f being the magnitude of the angular velocity of the leg just after its contact with the ground. Using the identity $1 = \cos^2\theta_0 + \sin^2\theta_0$, it follows that $(L_f)_z = -ml^2\omega_f$. Conservation of angular momentum yields $(L_f)_z = (L_i)_z$. This leads to:

$$\frac{\omega_f}{\omega_i} = \cos(2\theta_0). \tag{1.29}$$

(f) The person's kinetic energy just before the change of standing leg is $T_i = \frac{1}{2}ml^2\omega_i^2$, while just after the change of the standing leg it is $T_f = \frac{1}{2}ml^2\omega_f^2$. Using equation (1.29) the change of kinetic energy is $\Delta T = \frac{1}{2}ml^2\sin^2(2\theta_0)\,\omega_i^2$. Using equation (1.24) we obtain

$$\Delta T = \frac{1}{2}ml^2\sin^2(2\theta_0)\left[\frac{v_0^2}{l^2} + \frac{2g}{l}(1 - \cos\theta_0)\right]. \tag{1.30}$$

(g) In every step the person travels a distance of $2l\sin\theta_0$ and loses kinetic energy ΔT. The work per distance traveled that compensates for this loss of energy is $A' = \frac{\Delta T}{2l\sin\theta_0} = 59.9\,\frac{\text{J}}{\text{m}}$.

(h) To travel the distance of $s = 10\,\text{km}$ the person requires $E = A's = 599\,\text{kJ} = 143\,\text{kcal}$ of energy. The energy value of spinach per unit of mass is $w = \frac{28\,\text{kcal}}{100\,\text{g}} = 280\,\frac{\text{kcal}}{\text{kg}}$. Therefore, the person has to eat $m_s = \frac{E}{w} = 0.51\,\text{kg}$ of spinach.

We refer the reader interested in more detail about the physics of human walking to reference [1] and other references therein. The authors presented a modified version of this problem at the Serbian Physics Olympiad for high school students in 2018.

2

Machines

Problem 5   Microwave Oven

Why do microwaves not leak out of microwave ovens?
Why do microwave ovens contain a rotating plate?
Why is it difficult to cook an insect in a microwave oven?

Heating food in a microwave oven occurs due to the absorption of microwaves in the food. The main components of each microwave oven are a magnetron, a waveguide, and the oven itself. The magnetron consists of a cathode and an anode. High DC voltage (on the order of kV) is brought between these two electrodes. A magnet generates a magnetic field in the magnetron. Electrons move between the cathode and the anode in the presence of both the electric field and the magnetic field. The shape of these electrodes is such that a resonance occurs that leads to the creation of electromagnetic radiation the frequency of which is in the microwave range. The waveguide brings the microwaves from the magnetron to the oven where they heat the food. The walls of the oven are made of a metal that reflects electromagnetic radiation well.

(a) One of the oven walls is also the door of the oven and contains circular holes of diameter d. These holes should be large enough so that one can see the food heated in the oven. On the other hand, these holes should be also small enough to prevent the leakage of microwaves from the oven. Estimate the order of magnitude of d. The frequency of microwaves used in this oven is $f = 2.45\,\text{GHz}$.

(b) In this part of the problem we determine the absorption coefficient of microwaves in the food that is optimal in the sense that it heats the food in the best possible manner. If the absorption coefficient was too large, microwaves would be absorbed on the surface of the food and we would obtain food that is overheated on the surface and cold inside. On the other hand, if the absorption coefficient was too small, the food would not be heated at all. This discussion implies that there is an optimal absorption coefficient, which we determine as follows. Consider a monochromatic wave whose propagation direction is perpendicular to the surface of the food layer whose width is $b = 2.50\,\text{cm}$ and is located in the region $0 \le x \le b$. What is the position x_m in space where the food is least heated? Determine the absorption coefficient α_m that maximizes the absorption of microwaves at point x_m. The graph of the dependence of microwave absorption coefficient in water on frequency is shown in Figure 2.1. Determine the frequency of microwaves that maximizes the absorption at x_m.

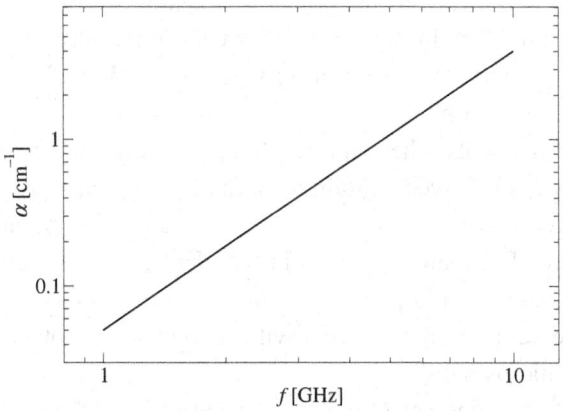

Figure 2.1 The dependence of microwave absorption coefficient in water on frequency

(c) In part (a) we estimated the dimension of the holes that prevents the leakage of microwaves outside the oven. Their leakage can be additionally minimized by the choice of the oven wall width, which leads to interference minimum in the transmission of microwaves through the wall. Consider a plane monochromatic wave propagating in the direction perpendicular to the oven wall. The refractive index of the wall material in the microwave range is $n_1 = 80.0$. In this part of the problem neglect for simplicity the absorption of microwaves in oven walls. Determine the positions of local minima in the dependence of the intensity of the transmitted microwaves on the wall width.

(d) Calculate the time needed to heat in the oven water of volume $V_w = 0.200\,\text{l}$ from temperature $t_1 = 20°\text{C}$ to $t_2 = 100°\text{C}$. The specific heat capacity of water is $c_w = 4,186\,\frac{\text{J}}{\text{kg·K}}$. Assume that the power of absorption of microwaves in water is $P = 500\,\text{W}$. The density of water is $\rho_w = 1,000\,\frac{\text{kg}}{\text{m}^3}$.

(e) The electric and magnetic fields in the oven are not homogeneous but form a standing wave as a consequence of wave reflections on the oven walls. Consider an oven in the shape of a cuboid whose sides are $L_x = L_y = 29.0\,\text{cm}$ and $L_z = 19.0\,\text{cm}$ in the region of space $0 \le x \le L_x$, $0 \le y \le L_y$, $0 \le z \le L_z$. The dependence of the magnitude of the electric field on spatial coordinates and time is given as

$$E(x,y,z,t) = \left| E_0 \sin\left(k_x x\right) \sin\left(k_y y\right) \sin\left(k_z z\right) \cos\left(\omega t - \delta\right) \right|, \qquad (2.1)$$

where $k_x = n_x \pi / L_x$, $k_y = n_y \pi / L_y$, $k_z = n_z \pi / L_z$, and $(n_x, n_y, n_z) = (3,4,3)$. E_0 is the amplitude of the electric field, ω is its angular frequency, and δ is the phase. An insect is located at the point $(x_0, y_0, z_0) = \left(\frac{8}{25}L_x, \frac{1}{3}L_y, \frac{1}{4}L_z\right)$ before the oven is turned on. The insect can move freely in the plane $z = \frac{1}{4}L_z$. To which point (x_f, y_f, z_f) should the insect go to be least affected by the microwaves when the oven turns on?

(f) Food located in points where the field is zero would not heat at all. For this reason, many microwaves contain a rotating plate that enables heating of all parts of the food. After the insect from part (e) goes to point (x_f, y_f, z_f), the plate is rotated. The plate is located in the plane $z = \frac{1}{4}L_z$ and it rotates around the point $(x_c, y_c, z_c) = \left(\frac{1}{2}L_x, \frac{1}{2}L_y, \frac{1}{4}L_z\right)$ at a frequency of $f_p = 0.100\,\text{s}^{-1}$. What should be the velocity of the insect with respect to the rotating plate to avoid the effect of microwaves?

(g) Quality factor is defined as $Q = 2\pi \frac{W_{em}}{W_g}$, where W_{em} is the total energy of electromagnetic radiation in the oven (averaged over one period of oscillations of the electromagnetic field), while W_g is the loss of electromagnetic energy in one oscillation. Assume that all electromagnetic energy losses heat the food and that the energy of the electric field in the oven is equal to the energy of the magnetic field. Determine the quality factor when water from part (d) is heated in the oven as described in part (e). Assume that $E_0 = 2.10 \cdot 10^4 \frac{\text{V}}{\text{m}}$. Reminder: The mean value of physical quantity x in time interval $(t_0, t_0 + T)$ is defined as $\langle x \rangle = \frac{1}{T} \int_{t_0}^{t_0+T} dt \cdot x(t)$.

(h) In most microwaves, a signal that turns off the magnetron is activated when a user opens the door. This stops the generation of microwaves. Nevertheless, it takes a certain time for the energy of electromagnetic radiation in the oven to decay to some small value. When will the energy of microwaves in the oven decay to $\frac{1}{1,000}$ part of its initial value?

In all parts of the problem you can use the following values of physical constants: the speed of light $c = 3.00 \cdot 10^8\,\text{ms}^{-1}$, vacuum permittivity $\varepsilon_0 = 8.85 \cdot 10^{-12}\,\frac{\text{F}}{\text{m}}$.

Solution of Problem 5

(a) The holes will prevent the leakage of microwaves if the wavelength λ_0 is much larger than the hole diameter $\lambda_0 \gg d$. The wavelength of the microwave is $\lambda_0 = c/f = 12.2$ cm, which leads to $d \ll 12.2$ cm. On the other hand, visible light will leave the oven if the hole diameter is much larger than the wavelength of visible light (which is around 500 nm), which is a condition that is certainly satisfied. Nevertheless, if the holes are too small, the intensity of the light that leaves the oven would be too small and it would not be possible to see the food from the oven. Therefore, we estimate that the optimal d is on the order of 1 cm.

(b) The intensity of the microwaves in the food layer depends on distance $I(x) = I_0 \exp(-\alpha x)$, where I_0 is the intensity of the incident microwave and α is the absorption coefficient of microwaves in food. Next, we consider a layer of food with cross-section S and small width Δx. In small time Δt the energy $\Delta W_{in} = I(x)S\Delta t$ enters this layer, while the energy $\Delta W_{out} = I(x+\Delta x)S\Delta t$ leaves it. Therefore, the energy absorbed by this layer in time Δt is $\Delta W = \Delta W_{in} - \Delta W_{out} = -\frac{dI}{dx}S\Delta x \Delta t$, which leads to $\Delta W = \alpha I_0 \exp(-\alpha x)\Delta V \Delta t$, where $\Delta V = S\Delta x$. The last expression implies that the absorbed energy decays exponentially with distance. Therefore, it is smallest for $x_m = b$. The energy at point $x_m = b$ is largest when the function $f(\alpha) = \alpha \exp(-\alpha b)$ has a maximum. The first derivative of this function is $f'(\alpha) = (1 - \alpha b)\exp(-\alpha b)$ and is equal to zero when $\alpha_m = 1/b = 0.400$ cm^{-1}. The expression for f' implies that $f' > 0$ when $\alpha < \alpha_m$ and $f' < 0$ when $\alpha > \alpha_m$. Consequently the function $f(\alpha)$ has a maximum when $\alpha = \alpha_m$. We read from Figure 2.1 that the frequency of microwaves is $f_m \approx 3.00$ GHz for $\alpha_m = 0.400$ cm^{-1}. This frequency is close to the frequency used in the microwave.

(c) We consider the following rays. Ray 1 passes through the wall without reflections. Ray 2 passes through the wall, it reflects at the outer surface of the wall, and then it reflects again at the inner surface of the wall and leaves the oven. Ray m exhibits $(m-1)$ reflections at the outer wall surface and $(m-1)$ reflections at the inner wall surface and finally leaves the oven. Rays 1, 2, ..., m, ... that leave the oven interfere. Minimum transmission occurs when the phase difference of rays $(m-1)$ and m is equal to an odd multiple of π. The phase difference of rays $(m-1)$ and m is $\Delta\phi = \frac{2\pi}{\lambda_0}n_1 \cdot 2d$, where d is the wall width and λ_0 is the wavelength of the microwave in a vacuum. From the condition $\Delta\phi = (2k+1)\pi$, where k is a nonnegative integer, it follows that $d = \left(\frac{1}{2}+k\right)\frac{\lambda_0}{2n_1} = \left(\frac{1}{2}+k\right)\cdot 765\,\mu\text{m}$.

(d) On one hand, the amount of heat absorbed in water in time t is $Q = Pt$. On the other hand, we have $Q = m_w c_w (t_2 - t_1)$, where $m_w = \rho_w V_w$ is the mass of water. From previous equations it follows that $t = \frac{\rho_w V_w c_w (t_2 - t_1)}{P} = 134$ s.

(e) We conclude from equation (2.1) that the magnitude of the electric field is zero in certain planes. Therefore, the insect should go to the nearest of these planes. The planes with zero electric field that are accessible to the insect are $x = \frac{L_x}{3}$, $x = \frac{2L_x}{3}$, $y = \frac{L_y}{4}$, $y = \frac{L_y}{2}$, and $y = \frac{3L_y}{4}$. Since the initial coordinates of the insect are $(x_0, y_0) = (\frac{8}{25}L_x, \frac{1}{3}L_y)$, the nearest plane is $x = \frac{L_x}{3}$. Therefore, the insect should go to point $(x_f, y_f, z_f) = (\frac{1}{3}L_x, \frac{1}{3}L_y, \frac{1}{4}L_z)$.

(f) The insect should remain in the same point of the laboratory reference frame. It should therefore move at a velocity equal in magnitude and opposite in direction to the velocity of the plate at the point where the insect is located. The magnitude of this velocity is $v = r \cdot 2\pi f_p$, where $r = \sqrt{(x_f - x_c)^2 + (y_f - y_c)^2}$, which leads to $v = 4.29 \frac{\text{cm}}{\text{s}}$.

(g) The energy density of the electric field is $w_e = \frac{1}{2}\varepsilon_0 E^2$, where

$$E(x,y,z,t)^2 = E_0^2 \sin^2(k_x x) \sin^2(k_y y) \sin^2(k_z z) \cos^2(\omega t - \delta) \tag{2.2}$$

Using the trigonometric identities $\sin^2 \alpha = \frac{1-\cos(2\alpha)}{2}$ and $\cos^2 \alpha = \frac{1+\cos(2\alpha)}{2}$, we conclude that the function $E(x,y,z,t)^2$ is periodic in the x-direction with the period of $b_x = \frac{\pi}{k_x}$, in the y-direction with the period of $b_y = \frac{\pi}{k_y}$, and in the z-direction with the period of $b_z = \frac{\pi}{k_z}$. We determine first the total energy of the electric field in the region of space whose shape is a cuboid with dimensions $b_x \times b_y \times b_z$. This energy is

$$W_b = \frac{1}{2}\varepsilon_0 E_0^2 \cos^2(\omega t - \delta) \int_0^{b_x} dx \sin^2(k_x x) \int_0^{b_y} dy \sin^2(k_y y) \int_0^{b_z} dz \sin^2(k_z z),$$
$$\tag{2.3}$$

which leads to

$$W_b = \frac{1}{2}\varepsilon_0 E_0^2 \cos^2(\omega t - \delta) \frac{1}{8} b_x b_y b_z. \tag{2.4}$$

Using the definition of mean value given in the problem (where $T = \frac{2\pi}{\omega}$ is the period of oscillations) we find

$$\langle \cos^2(\omega t - \delta) \rangle = \frac{1}{T} \int_0^T \cos^2(\omega t - \delta) = \frac{1}{2}. \tag{2.5}$$

This leads to

$$\langle W_b \rangle = \frac{1}{32}\varepsilon_0 E_0^2 b_x b_y b_z. \tag{2.6}$$

From the proportion $\frac{\langle W_b \rangle}{W_e} = \frac{b_x b_y b_z}{L_x L_y L_z}$, where W_e is the total energy in the oven averaged over one oscillation, we find $W_e = \frac{1}{32}\varepsilon_0 E_0^2 L_x L_y L_z$. Since the energy of the electric field is equal to the energy of the magnetic field, the total energy of the electromagnetic field is $W_{em} = 2W_e = \frac{1}{16}\varepsilon_0 E_0^2 L_x L_y L_z$. Next we have $W_g = P \cdot T$. Consequently the quality factor is $Q = \frac{1}{16}\varepsilon_0 E_0^2 L_x L_y L_z \omega / P = 120$.

(h) The energy of the electromagnetic wave reduces by $\Delta W_{em} = 2\pi W_{em}/Q$ in one oscillation. Since $\Delta W_{em} \ll W_{em}$, we can use the approximation $\frac{dW_{em}}{dt} = -\frac{\Delta W_{em}}{T}$, where $T = \frac{2\pi}{\omega}$ is the period of the wave. This leads to $\frac{dW_{em}}{dt} = -\frac{\omega}{Q}W_{em}$ and therefore $W_{em}(t) = W_{em}(0)\exp\left(-\frac{\omega}{Q}t\right)$. The time t_X it takes for the field to decay to $\frac{1}{1,000}$ part of the initial value can be found from the condition $\frac{1}{1,000} = \exp\left(-\frac{\omega}{Q}t_X\right)$. Therefore, $t_X = \frac{3Q}{\omega}\ln 10 = 54\,\text{ns}$. Since it is not possible to open the door that fast, the microwaves cannot leak outside the oven during the opening of the door.

We refer the reader interested in the physics of microwave ovens to reference [36].

Problem 6   Defibrillator

How does an electric shock of the heart help in the case of cardiac arrhythmia?

Ventricular fibrillation is the cause of many deaths. It is a type of cardiac arrhythmia that occurs due to disorganized contraction of the muscles in the heart chambers. The heart pumps only a small amount of blood in this condition. Therefore, the brain remains without blood and loss of consciousness occurs only a few seconds after the start of ventricular fibrillation. Brain death then occurs after a few minutes. This can be prevented by the application of an electric shock to the heart. A human feels an electric shock if the current through the body is larger than $I_{min} = 1\,\text{mA}$.

(a) A person has accidentally touched the electrical heater of a toaster with a metal knife. Calculate the minimal resistance of the system "metal knife – human body" that will prevent a person from feeling the electric current. The toaster is connected to an AC power system the effective voltage of which is $U_{eff} = 220\,\text{V}$.

The electric shock of a patient with ventricular fibrillation is caused using a medical device called a defibrillator. Its electrical scheme is shown in Figure 2.2. A defibrillator consists of an AC/DC converter and a capacitor that is first charged to a high voltage (the switch in position 1) and then discharged through the patient's body (the switch in position 2). The electric shock instantaneously paralyzes the heart muscle, which allows the heart to return to its regular condition.

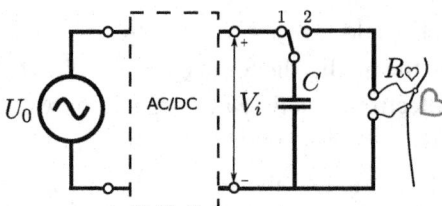

Figure 2.2 Electric scheme of a defibrillator

We consider an AC/DC voltage converter in this part of the problem. Its electric scheme is shown in Figure 2.3. The converter consists of an ideal transformer connected to an AC power system through a primary the winding of which has $N_0 = 200$ turns and a secondary the winding of which has $N_1 = 2,000$ turns, four ideal diodes D, the capacitor $C_p = 100\,\mu\text{F}$, and the resistor $R_p = 3.00\,\text{k}\Omega$. The effective voltage from the AC power system is $U_{\text{eff}} = 220\,\text{V}$, while its frequency is $\nu = 50.0\,\text{Hz}$. An ideal diode is an electronic component that allows the flow of electric current without resistance in one direction only (when it is directly polarized), while its resistance is infinite when it is inversely polarized.

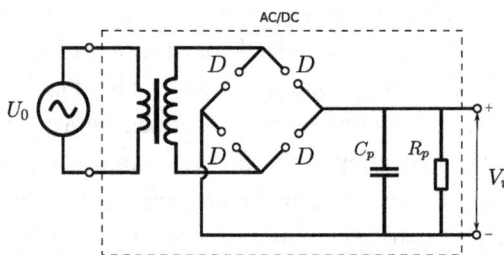

Figure 2.3 Electric scheme of the AC/DC voltage converter

(b) How should the diodes be connected to the circuit so that the polarity of the output voltage V_i is as shown in Figure 2.3?

(c) Plot the time dependence of the output voltage when the capacitor C_p is not connected to the electric circuit. The dependence of the voltage of the power system on time is sinusoidal $u(t) = U_0 \sin(2\pi\nu t)$.

(d) Plot the time dependence of the output voltage of the converter when the capacitor C_p is connected to the electric circuit. The capacitor C_p is discharged at time $t = 0$, while the dependence of the voltage of the power system on time is sinusoidal $u(t) = U_0 \sin(2\pi\nu t)$. What is the role of the capacitor in this converter?

(e) Determine the minimal V_{min} and maximal V_{max} voltage at the output of the converter. Calculate the error incurred if we assume that the voltage at the output of the converter is constant $V_i = \frac{1}{2}(V_{max} + V_{min})$.

Next, we consider how the defibrillator causes the electrical shock of the heart. Assume that the output voltage of the converter has a constant value V_i. After the defibrillator is connected to the patient's body, the switch in Figure 2.2 is moved from position 1 to position 2. The current then flows through the patient's body. The time interval during which the current larger than I_{min} flows through the body should be smaller than $\Delta t_{max} = 30.0\,\mathrm{ms}$ in order to avoid causing burns on the patient's skin. The electrical resistance of the heart during electric shock is constant and takes the values in the interval from $R_\heartsuit^{min} = 50\,\Omega$ to $R_\heartsuit^{max} = 150\,\Omega$, depending on the state of the patient, the type of defibrillator, etc.

(f) Calculate the maximal capacitance $C = C_{max}$ of the capacitor in the defibrillator. Calculate the energy of that capacitor when it is charged.
(g) Plot the time dependence of the charge that passes through the heart when $R_\heartsuit = 100\,\Omega$ and $C = C_{max}$.

Biphasic defibrillators have recently become popular. They are made in such a way that the direction of the current changes during the electric shock. This is achieved using two capacitors. The probability of burns on the patient's skin is decreased this way. In the first phase of the shock the capacitor C_1 is discharging during a short time interval. The polarity of the voltage is then suddenly changed, and in the second phase the capacitor C_2 starts discharging. The total energy transferred to the patient during a electric shock that lasts $\Delta t = 18\,\mathrm{ms}$ is $E = 180\,\mathrm{J}$. The time dependence of the voltage on the capacitors during the use of one such defibrillator is shown in Figure 2.4.

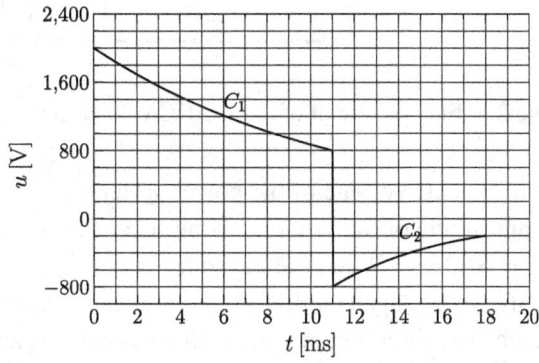

Figure 2.4 Biphasic defibrillator

(h) Calculate the capacitance of the capacitors C_1 and C_2 and the resistance $R'_\heartsuit$.
(i) Calculate the mean electric current that passes through the patient's body in each of the phases of the electric shock.
(j) What fraction of the total energy is released in each of the phases of the electric shock?

Solution of Problem 6

(a) The minimal electrical resistance of the system "metal knife – human body" should be $R_{min} = \frac{U_0}{I_{min}}$, where $U_0 = U_{eff}\sqrt{2} = 311$ V is the amplitude of the power system voltage, which leads to $R_{min} = 0.311\,M\Omega$. Electrical resistance of the human skin makes the largest contribution to the total electrical resistance of the body. When the skin is dry, the resistance between two hands can be on the order of $M\Omega$. It is therefore very dangerous to become a part of an electric circuit in this way.

(b) The output voltage V_i has the desired polarity if a DC electric current flows through the resistor. For this reason, the diodes should be connected as shown in Figure 2.5. The diodes then conduct in pairs. The diodes D_1 and D_2 conduct in one half period, while the diodes D_3 and D_4 conduct in the other half period. The direction of current through certain parts of the circuit for each of the polarities of the input AC voltage is shown in solid and dashed arrows.

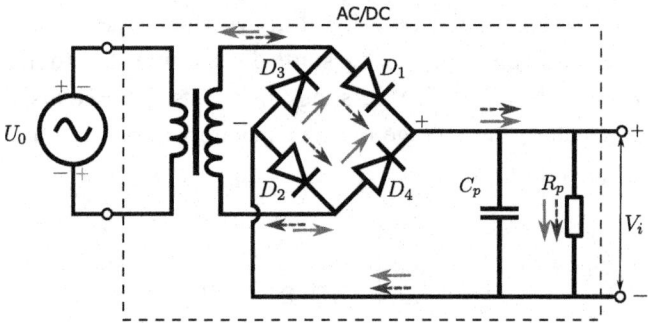

Figure 2.5 Electric circuit of the AC/DC voltage converter

(c) The amplitude of the AC voltage of the power system is $U_0 = U_e\sqrt{2} = 311$ V. Therefore, the amplitude of the voltage on the secondary of the transformer is $U_1 = N_1U_0/N_0 = 3.11$ kV. When the capacitor is not connected to the circuit the voltage on the secondary is equal to the voltage on the resistor, which is the output voltage. The direction of current in the power system changes every $T_{1/2} = T/2 = 1/(2\nu) = 10.0\,ms$, which is the period of the output voltage V_i. The graph with time dependence of the output voltage is shown in Figure 2.6.

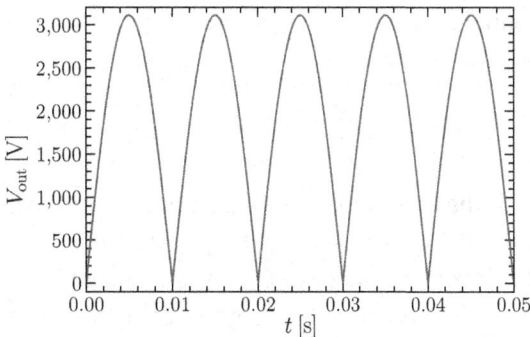

Figure 2.6 Time dependence of the output voltage

(d) The diodes D_1 and D_2 conduct in the first half period and the capacitor is charged up to maximal voltage $V_{\max} = U_1$, which is reached at time $t_1 = \frac{T_{1/2}}{2} = \frac{1}{4v}$. The voltage on the secondary then decreases and the diode D_1 stops conducting. The capacitor then discharges through the resistor R_p. The capacitor discharges until time t_2 when the voltages on the capacitor and the secondary become equal again. The diodes D_3 and D_4 start conducting after that, which leads to the charging process of the capacitor up to maximal voltage that is reached at time $t_3 = t_1 + T_{1/2} = \frac{3}{4v}$. The whole process then repeats with period $T_{1/2}$.

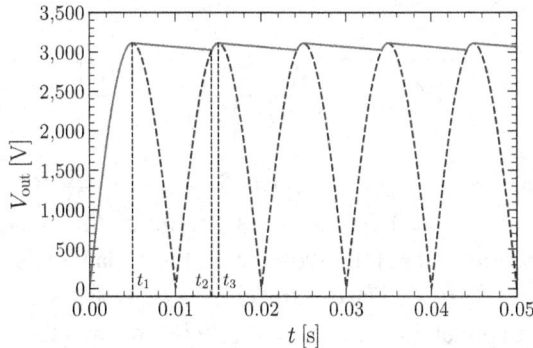

Figure 2.7 Time dependence of the output voltage when the capacitor is connected (solid line) and when it is not connected (dashed line)

During discharging of the capacitor (when $t_1 \leq t \leq t_2$) the voltages on the capacitor u_{C_p} and the resistor u_{R_p} satisfy the, condition $u_{C_p} + u_{R_p} = 0$, which leads to

$$\frac{\mathrm{d}}{\mathrm{d}t} u_{C_p} + \frac{1}{R_p C_p} u_{C_p} = 0 . \tag{2.7}$$

This is a differential equation with separated variables. It is solved by integration, which leads to

$$u_{C_p}(t) = u_{C_p}(t_1) e^{-\frac{t-t_1}{\tau}} = V_{\max} e^{-\frac{t-t_1}{\tau}}, \tag{2.8}$$

where $\tau = R_p C_p$ is the time constant of the circuit. Since the discharging time of the capacitor is smaller than $T_{1/2}$ and $\tau \gg T_{1/2}$, we find $u_{C_p}(t) \approx V_{\max}(1 - \frac{t-t_1}{\tau})$. The graph of the time dependence of the output voltage is shown in Figure 2.7. If the time constant of the circuit is large in comparison to the period of the input AC voltage, the voltage on the resistor changes only slightly during the discharge of the capacitor. This way the variations of the output voltage are also smaller. Since the time constant of the circuit depends on the capacitance, the appropriate choice of the capacitor can significantly reduce the variations of the output voltage.

(e) The maximal voltage at the output of the converter is $V_{\max} = 3.11\,\text{kV}$, while the minimal voltage is obtained by solving the transcendental equation

$$V_{\min} = |U_1 \sin(2\pi v t_2)| = U_1 \left(1 - \frac{t_2 - \frac{1}{2}T_{1/2}}{\tau} \right) \tag{2.9}$$

with respect to t_2, where $10.0\,\text{ms} < t_2 < 15.0\,\text{ms}$. We obtain $t_2 = 14.2\,\text{ms}$ and $V_{\min} = 3.02\,\text{kV}$. The relative error that we make assuming that the output voltage at time t is $V_i = \frac{1}{2}(V_{\max} + V_{\min})$ reads

$$\delta V_i(t) = \frac{|\frac{1}{2}(V_{\max} + V_{\min}) - V_i(t)|}{V_i(t)}. \tag{2.10}$$

It is maximal for $V_i(t) = V_{\min}$ and reads $(\delta V_i)_{\max} = \frac{V_{\max} - V_{\min}}{2V_{\min}} \cdot 100\% = 1.58\%$

(f) The electric shock of the heart is the discharge of the capacitor in the $R_\heartsuit C$ circuit. As shown in part (c), the voltage on the capacitor is given as $u_C(t) = V_i e^{-t/\tau_\heartsuit}$, where $\tau_\heartsuit = R_\heartsuit C$. The electric current that passes through the patient's body decreases exponentially as $i(t) = V_i e^{-t/\tau_\heartsuit}/R_\heartsuit$, and the electric shock finishes when it reaches the value of $I_{\min}$. Consequently the duration of the electric shock is $\Delta t = R_\heartsuit C \ln\left(\frac{V_i}{R_\heartsuit I_{\min}}\right)$. To calculate the maximal capacitance, we need first to investigate the function $\Delta t(R_\heartsuit)$ in the interval of arguments $[R_\heartsuit^{\min}, R_\heartsuit^{\max}]$. The first derivative of this function is $(\Delta t)'(R_\heartsuit) = C\left[\ln\left(\frac{V_i}{R_\heartsuit I_{\min}}\right) - 1\right]$. It is equal to zero when $R_\heartsuit^{\text{ext}} = \frac{V_i}{e I_{\min}}$. From the expression for $(\Delta t)'(R_\heartsuit)$, we see that for $R < R_\heartsuit^{\text{ext}}$ we have $(\Delta t)'(R_\heartsuit) > 0$, while for $R > R_\heartsuit^{\text{ext}}$ we have $(\Delta t)'(R_\heartsuit) < 0$. The function $\Delta t(R_\heartsuit)$ therefore exhibits a maximum for

$R = R_\heartsuit^{\text{ext}}$. Since $R_\heartsuit^{\text{ext}} = 1.13\,\text{M}\Omega > R_\heartsuit^{\text{max}}$, the function $\Delta t(R_\heartsuit)$ is monotonously increasing in the interval $[R_\heartsuit^{\text{min}}, R_\heartsuit^{\text{max}}]$. The maximal value of the capacitance is then

$$C_{\max} = \frac{\Delta t_{\max}}{R_\heartsuit^{\text{max}} \ln\left(\frac{V_i}{R_\heartsuit^{\text{max}} I_{\min}}\right)} = 20.2\,\mu\text{F}. \tag{2.11}$$

The energy of the charged capacitor is $E = \frac{1}{2} C_{\max} V_i^2 = 94.6\,\text{J}$.

(g) The voltage on the capacitor is $U(t) = V_i e^{-\frac{t}{R_\heartsuit C_{\max}}}$, while the charge is $q(t) = C_{\max} U(t) = C_{\max} V_i e^{-\frac{t}{R_\heartsuit C_{\max}}}$. Consequently the amount of charge that flowed through the heart is $\Delta q = C_{\max} V_i (1 - e^{-\frac{t}{R_\heartsuit C_{\max}}})$. The corresponding graph is given in Figure 2.8.

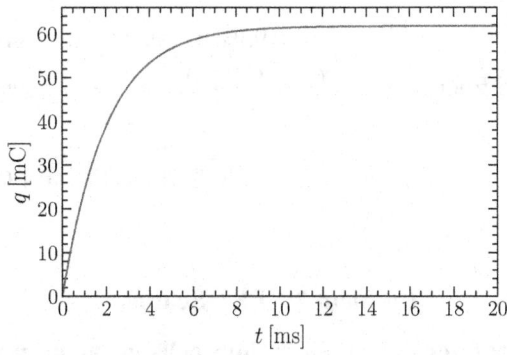

Figure 2.8 Time dependence of the amount of charge that passes through the heart

(h) The graph given in the text of the problem implies that the capacitor C_1 is discharged from the voltage $V_1 = 2{,}000\,\text{V}$ to $V_2 = 800\,\text{V}$ during the time interval $\Delta t_1 = 11\,\text{ms}$ in the first phase. It holds that

$$V_2 = V_1 e^{-\frac{\Delta t_1}{R'_\heartsuit C_1}}. \tag{2.12}$$

In the second phase, the capacitor C_2 discharges from the voltage V_2 to $V_3 = 200\,\text{V}$ during the time interval $\Delta t_2 = 7\,\text{ms}$. It therefore holds that

$$V_3 = V_2 e^{-\frac{\Delta t_2}{R'_\heartsuit C_2}}. \tag{2.13}$$

The total energy released E is

$$E = \frac{1}{2} C_1 (V_1^2 - V_2^2) + \frac{1}{2} C_2 (V_2^2 - V_3^2). \tag{2.14}$$

By solving the system of equations (2.12)–(2.14), we obtain

$$C_1 = \frac{2E\Delta t_1 \ln \frac{V_2}{V_3}}{(V_1^2 - V_2^2)\Delta t_1 \ln \frac{V_2}{V_3} + (V_2^2 - V_3^2)\Delta t_2 \ln \frac{V_1}{V_2}} = 99.7\,\mu F,$$

$$C_2 = \frac{2E\Delta t_2 \ln \frac{V_1}{V_2}}{(V_1^2 - V_2^2)\Delta t_1 \ln \frac{V_2}{V_3} + (V_2^2 - V_3^2)\Delta t_2 \ln \frac{V_1}{V_2}} = 41.9\,\mu F,$$

$$R_\heartsuit' = \frac{(V_1^2 - V_2^2)\Delta t_1 \ln \frac{V_2}{V_3} + (V_2^2 - V_3^2)\Delta t_2 \ln \frac{V_1}{V_2}}{2E \ln \frac{V_1}{V_2} \ln \frac{V_2}{V_3}} = 120\,\Omega.$$

(i) The mean electric current that passes through the patient's body in the first phase is $I_1 = \frac{Q_1 - Q_2}{\Delta t_1} = \frac{C_1(V_1 - V_2)}{\Delta t_1} = 10.9\,A$, while it is $I_2 = \frac{Q_2 - Q_3}{\Delta t_2} = \frac{C_2(V_2 - V_3)}{\Delta t_2} = 3.59\,A$ in the second phase.

(j) The fraction $\eta_1 = \frac{E_1}{E} = \frac{C_1(V_1^2 - V_2^2)}{2E} = 93\%$ of total energy is released in the first phase, while the fraction $\eta_2 = \frac{E_2}{E} = \frac{C_2(V_2^2 - V_3^2)}{2E} = 7\%$ is released in the second phase.

The authors presented this problem at the Serbian Physics Olympiad for high school students in 2017.

Problem 7   PET Scanner

How does a PET scanner detect anomalous cells in the human body?
Why does a patient examined by PET drink a solution that contains glucose?

Scanning based on positron emission tomography (PET) is a diagnostic technique used in medicine. A patient treated by this technique first drinks a solution that contains glucose in which oxygen nuclei are replaced by nuclei of the radioactive isotope $^{18}_{9}F$. This isotope emits positrons in β^+ decay. Each positron in the patient's body annihilates with an electron leading to the creation of two gamma ray photons. By detecting these photons one can detect where in the body $^{18}_{9}F$ nuclei are present in larger quantities, such as the cells in the brain and cancer cells.

(a) During preparation of the solution $^{18}_{8}O$ nuclei are bombed by fast protons. This leads to the creation of $^{18}_{9}F$ nuclei and neutrons. Write down the equation of this nuclear reaction.

(b) Write down the equation that describes the β^+ decay of the $^{18}_{9}F$ nuclei.

(c) Determine the maximal kinetic energy of the positrons T_{max} created in the β^+ decay of the $^{18}_{9}F$ nuclei that are at rest.

(d) The positron created in the β^+ decay decelerates in the human body and stops at a distance on the order of a millimeter. It then annihilates with an electron, which leads to the creation of two gamma ray photons. Determine the energies of each of these photons.

(e) To determine the position in the human body where annihilation occurred, a PET scanner contains the ring shown in Figure 2.9 with a large number of detectors that register pairs of gamma ray photons that arrive almost simultaneously. The process of annihilation at point P (see Figure 2.9) led to the counts on the detectors A and B, where the count on the detector A happened $\Delta t = 0.310$ ns before the count on the detector B. What is the distance between the point where the annihilation occurred and the midpoint of the line AB? Assume that the speed of gamma ray photons is equal to the speed of light in a vacuum.

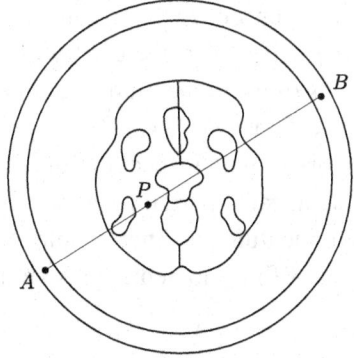

Figure 2.9 Scheme of the ring-shaped detector of the PET scanner and the brain of the patient

(f) The patient received a solution of glucose that contains $^{18}_{9}F$ nuclei. The initial gamma activity of this solution is $A_\gamma = 90.0$ MBq. Determine the initial β^+ activity of this solution.

(g) Find the initial number of $^{18}_{9}F$ nuclei in the solution.

(h) Determine the total energy absorbed in the patient's body before all $^{18}_{9}F$ nuclei decay. Assume that the kinetic energy of the positron is on average equal to $T_{\mathrm{sr}} = \frac{1}{3}T_{\max}$, as well as that $\alpha = 40.0\%$ of gamma ray photons that are created by annihilation get absorbed in the patient's body. Assume that all $^{18}_{9}F$ nuclei decay in the patient's body.

You can use the following numerical values of physical quantities and constants. The mass of the $^{18}_{9}F$ nuclei is $m_F = 18.000937 \cdot u$, the mass of the $^{18}_{8}O$ nuclei is $m_O = 17.999160 \cdot u$, and the mass of the positron is equal to the mass of the electron

$m_{e^+} = m_{e^-} = 0.000548756 \cdot u$. Assume that the mass of the neutrino is $m_v = 0$. The atomic unit of mass is $u = 931.49409 \frac{\text{MeV}}{c^2}$. The half-life of the $^{18}_9\text{F}$ nuclei for the β^+ process is $T_{1/2} = 109.8\,\text{min}$. The speed of light in a vacuum is $c = 3.00 \cdot 10^8 \frac{\text{m}}{\text{s}}$, and the elementary charge is $e = 1.60 \cdot 10^{-19}\,\text{C}$.

Solution of Problem 7

(a) The equation of the reaction is

$$^{18}_8\text{O} + ^1_1 p \longrightarrow ^{18}_9\text{F} + ^1_0 n. \tag{2.15}$$

(b) In β^+ decay the $^{18}_9\text{F}$ nucleus transforms into $^{18}_8\text{O}$ with the emission of the positron and the neutrino. The equation of the reaction therefore reads

$$^{18}_9\text{F} \longrightarrow ^{18}_8\text{O} + ^0_1 e^+ + v. \tag{2.16}$$

(c) For each of the particles, we introduce the four-momentum defined as the four-vector $P = \left(\frac{E}{c}, p_x, p_y, p_z\right)$, where E is the energy of the particle, while p_x, p_y, and p_z are the components of its momentum. We also define the scalar product of two four-vectors P_1 and P_2 as $(P_1, P_2) = \frac{E_1 E_2}{c^2} - \vec{p}_1 \cdot \vec{p}_2$. It can be shown that this scalar product is invariant with respect to Lorentz transformation – that is, that it has the same value in each inertial reference frame.

Since conservation of momentum and energy holds for the nuclear reaction, it follows that conservation of four-momentum also holds, which leads to

$$P_F = P_{e^+} + P_O + P_v. \tag{2.17}$$

We next have $P_F - P_{e^+} = P_O + P_v$; that is, $(P_F - P_{e^+}, P_F - P_{e^+}) = (P_O + P_v, P_O + P_v)$, which leads to

$$(P_F, P_F) + (P_{e^+}, P_{e^+}) - 2 \cdot (P_F, P_{e^+}) = (P_O + P_v, P_O + P_v). \tag{2.18}$$

Using $(P_F, P_F) = m_F^2 c^2$, $(P_{e^+}, P_{e^+}) = m_{e^+}^2 c^2$ and $(P_F, P_{e^+}) = m_F \left(T + m_{e^+} c^2\right)$ (where T is the kinetic energy of the positron), we obtain

$$T = \frac{m_F^2 c^2 + m_{e^+}^2 c^2 - (P_O + P_v, P_O + P_v)}{2m_F} - m_{e^+} c^2. \tag{2.19}$$

This expression is maximal when the term $(P_O + P_v, P_O + P_v)$ is minimal. Since this term has the same value in all inertial reference frames, we calculate it in the reference frame where it holds that $\vec{p}_O + \vec{p}_v = 0$. In this reference frame we have $(P_O + P_v, P_O + P_v) = m_O c + m_v c + \frac{T_O + T_v}{c}$ and this expression is minimal when $\vec{p}_O = \vec{p}_v = 0$. Therefore, we finally obtain

$$T_{\text{max}} = \frac{m_F^2 + m_{e^+}^2 - m_O^2}{2m_F} c^2 - m_{e^+} c^2 = 1.14403\,\text{MeV}. \tag{2.20}$$

(d) Since the positron and the electron were at rest before annihilation, conservation of momentum reads $\vec{p}_1 + \vec{p}_2 = 0$, where $\vec{p}_1$ and $\vec{p}_2$ are the momenta of gamma photons after annihilation. This leads to $p_1 = p_2$. Therefore, emitted gamma photons have the same frequency but they move in opposite directions. Conservation of energy then yields $m_{e^+}c^2 + m_{e^-}c^2 = 2E_\gamma$ – that is,

$$E_\gamma = \frac{1}{2}(m_{e^+} + m_{e^-})c^2 = 0.511163\,\text{MeV}. \tag{2.21}$$

(e) Consider the annihilation process that took place in point P at time t. We then have $c(t_A - t) = PA$ and $c(t_B - t) = PB$, where t_A and t_B are the moments of detection of the photons in points A and B. It holds that $PB - PA = 2d$, where d is the distance between point P and the midpoint of line AB. Using $\Delta t = t_B - t_A$, previous equations give

$$d = \frac{c}{2}\Delta t = 4.65\,\text{cm}. \tag{2.22}$$

(f) One positron is created in each β^+ decay, which then gives two gamma photons after annihilation with an electron. For this reason, γ activity of the solution is twice larger than the β^+ activity – that is, $A_{\beta^+} = \frac{1}{2}A_\gamma = 45.0\,\text{MBq}$.

(g) Initial activity of the solution and initial number of $^{18}_9\text{F}$ nuclei N are related as $A_{\beta^+} = \frac{\ln 2}{T_{1/2}}N$, which leads to $N = \frac{A_{\beta^+}T_{1/2}}{\ln 2} = 4.28 \cdot 10^{11}$.

(h) The amount of energy absorbed in the patient's body is $Q = \alpha N_\gamma E_\gamma + NT_{\text{sr}}$, where $N_\gamma = 2N$ is the number of gamma photons created in the process of annihilation. This leads to $Q = 2\alpha N E_\gamma + \frac{1}{3}NT_{\text{max}} = 0.0541\,\text{J}$.

This problem was created by modifying the problem given at a physics olympiad in Denmark in 2009.

Problem 8   Induction Motor

What makes the induction motor rotate?

The induction motor is one of most widely used types of electric motors. It is used in dishwashers, ventilators, pumps, etc. It was patented by Nikola Tesla in 1888. An induction motor consists of the stator, which is a static part powered by the source of alternating current, and the rotor, which rotates due to the influence of the magnetic field of the stator. The magnitude of the magnetic field is constant in the region where the rotor is located, while the direction of the magnetic field rotates at a constant angular velocity around the axis of the rotor.

We first consider how one can produce such a magnetic field.

(a) A cylinder-shaped coil of radius a is located in the region $z_1 \leq z \leq z_2$, where the z-axis coincides with the axis of the cylinder, as shown in Figure 2.10. A current of surface density i_l runs through the coil. The assumed direction of this current is shown in Figure 2.10. Determine the vector of the magnetic field at an arbitrary point on the axis of the coil.

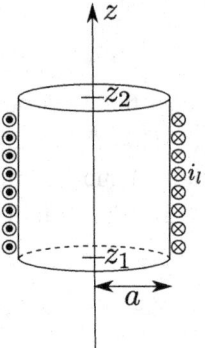

Figure 2.10 A coil in the shape of the cylinder of radius a that is located in the region $z_1 \leq z \leq z_2$

(b) Four identical coils of length l and radius a are positioned as shown in Figure 2.11. The distance between point A and the end of each of the coils is d. An alternating current the surface density of which depends on time as $i_m(t) = i_0 \cos\left[\omega_0 t - (m-1)\frac{\pi}{2}\right]$ is flowing through coil m. The assumed directions of current flow are shown in the figure, while i_0 and ω_0 are known constants. Determine the vector of the magnetic field at point A at time t.

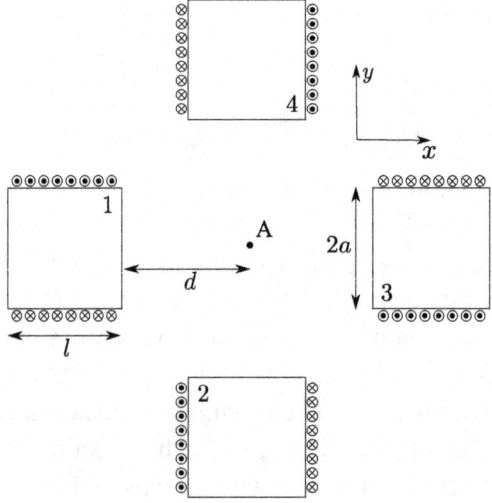

Figure 2.11 Four identical coils of length l and radius a

We assume next that the stator produces the magnetic field of magnitude B_S, whose direction is in the xy-plane and rotates around the z-axis at an angular velocity ω_0, as shown in Figure 2.12. The rotor consists of a rectangular metal contour connected to the axle, which is made of insulating material. The axle is in the z-direction and the rotor can rotate around this axis. The resistance of the metal contour is R, its inductance is L, and its area is A.

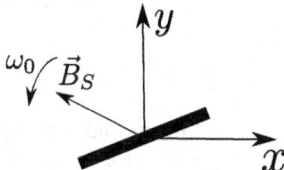

Figure 2.12 Projection of the contour to the xy-plane. The direction of the magnetic field at a certain moment of time is shown in the figure

(c) Determine the angular frequency and the amplitude of the current that flows through the rotor. The following quantities may appear in these expressions: B_S, A, ω_0, R, L, and angular velocity of the rotor $\omega < \omega_0$.

(d) Determine the mean value of the torque M_{av} that acts on the rotor. The final expression should contain the following quantities: B_S, A, ω_0, R, L, and $s = \frac{\omega_0 - \omega}{\omega_0}$.

(e) For what s from the interval $[0, 1]$ is M_{av} maximal?

We assume next that the torque acting on the rotor due to the effect of the magnetic field is equal to M_{av} at each moment of time. Due to the effects of air resistance and friction in the axle, a torque of magnitude $M_r = K\omega$ also acts on the rotor, where K is the resistance coefficient. Numerical values of relevant quantities are as follows: $\omega_0 = 100 \cdot \pi$ Hz, $R = 3.6\,\Omega$, $L = 0.12$ H, $A = 104$ cm^2, $B_S = 135$ mT, and $s \in [0, 1]$.

(f) Determine possible values of s if $K = 0.75 \cdot 10^{-8}\ \frac{\text{N·m·s}}{\text{rad}}$.

(g) Determine possible values of s if $K = 1.25 \cdot 10^{-8}\ \frac{\text{N·m·s}}{\text{rad}}$.

(h) Which of the solutions in parts (f) and (g) are stable?

Solution of Problem 8

(a) From the symmetry of the problem, we conclude that it is only the z-component of the magnetic field that has nonzero values on the axis of the coil. To determine this component, we divide the coil into small parts of length dz and we add the contributions of each of these parts. From the formula for the magnetic

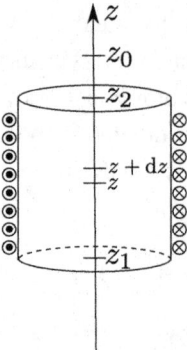

Figure 2.13 The coil in the shape of the cylinder of radius a that is located in the region $z_1 \leq z \leq z_2$. The part of the coil that is in the region $(z, z + dz)$ is shown in the figure, as well as the point z_0 in which the magnetic field is calculated

field of a circular contour on its axis, we find that the contribution of the part of the coil which is located in the region $(z, z + dz)$ to the magnetic field at point z_0 (see Figure 2.13) is equal to

$$dB_z = \frac{\mu_0 a^2 i_l dz}{2\left[(z - z_0)^2 + a^2\right]^{3/2}}. \tag{2.23}$$

The total z-component of the magnetic field at point z_0 is then

$$B_z = \frac{\mu_0 i_l a^2}{2} \int_{z_1}^{z_2} \frac{dz}{\left[(z - z_0)^2 + a^2\right]^{3/2}}. \tag{2.24}$$

Next we introduce the replacement $z = z_0 - a\tan\varphi$, which leads to $dz = -a\frac{d\varphi}{\cos^2\varphi}$. Using the identity $\tan^2\varphi + 1 = \frac{1}{\cos^2\varphi}$ it follows that

$$B_z = -\frac{\mu_0 i_l}{2} \int_{\varphi_1}^{\varphi_2} \cos\varphi \, d\varphi, \tag{2.25}$$

where the angles φ_1 and φ_2 are determined by the relation

$$\sin\varphi_i = \frac{z_0 - z_i}{\sqrt{(z_0 - z_i)^2 + a^2}}. \tag{2.26}$$

By solving the integral we obtain the final expression for the z-component of the magnetic field at the point z_0 on the axis of the coil:

$$B_z = \frac{\mu_0 i_l}{2} \left(\frac{z_0 - z_1}{\sqrt{(z_0 - z_1)^2 + a^2}} - \frac{z_0 - z_2}{\sqrt{(z_0 - z_2)^2 + a^2}} \right). \tag{2.27}$$

(b) Using the solution of part (a), we find that the contributions of coils to the magnetic field at point A are respectively:

$$\vec{B}_1 = ci_1\vec{e}_x, \tag{2.28}$$

$$\vec{B}_2 = ci_2\vec{e}_y, \tag{2.29}$$

$$\vec{B}_3 = -ci_3\vec{e}_x, \tag{2.30}$$

$$\vec{B}_4 = -ci_4\vec{e}_y, \tag{2.31}$$

where

$$c = \frac{\mu_0}{2}\left(\frac{l+d}{\sqrt{(l+d)^2+a^2}} - \frac{d}{\sqrt{d^2+a^2}}\right). \tag{2.32}$$

The total vector of the magnetic field at point A is

$$\vec{B} = c(i_1 - i_3)\vec{e}_x + c(i_2 - i_4)\vec{e}_y. \tag{2.33}$$

Using $i_1 = -i_3$ and $i_2 = -i_4$ it follows that

$$\vec{B} = 2ci_0\cos(\omega_0 t)\vec{e}_x + 2ci_0\sin(\omega_0 t)\vec{e}_y. \tag{2.34}$$

We see from this expression that the magnitude of the magnetic field at point A is equal to $2ci_0$ at each moment of time, while the vector of the magnetic field rotates at an angular velocity of ω_0 around the z-axis.

(c) The flux of the magnetic field through the contour is

$$\Phi = ABs\cos(\theta_S - \theta_n), \tag{2.35}$$

where θ_S is the angle between the direction of $\vec{B}_S$ and the x-axis, while θ_n is the angle between the direction of $\vec{n}$ and the x-axis, where $\vec{n}$ is a unit vector perpendicular to the plane of the contour; see Figure 2.14.

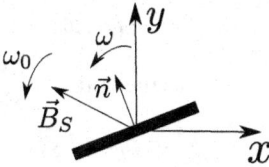

Figure 2.14 Projection of the contour to the xy-plane. The direction of the magnetic field and the vector perpendicular to the plane of the contour in a certain moment of time are shown in the figure

The induced electromotive force in the contour is obtained from Faraday's law

$$\varepsilon_i = -\frac{d\Phi}{dt} - L\frac{di_C}{dt},$$ (2.36)

where i_C is the current in the contour. From Ohm's law

$$\varepsilon_i = Ri_C,$$ (2.37)

it follows that

$$B_S A \left(\frac{d\theta_S}{dt} - \frac{d\theta_n}{dt} \right) \sin(\theta_S - \theta_n) = Ri_C + L\frac{di_C}{dt}.$$ (2.38)

The angles θ_S and θ_n depend on time as $\theta_S = \omega_0 t + \theta_{S0}$ and $\theta_n = \omega t + \theta_{n0}$, which leads to

$$B_S A (\omega_0 - \omega) \sin[(\omega_0 - \omega)t + \theta_{S0} - \theta_{n0}] = Ri_C + L\frac{di_C}{dt}.$$ (2.39)

The quantity on the left-hand side of the previous equation is a sinusoidal function of time with angular frequency $\omega_0 - \omega$. Consequently the angular frequency of the current must have the same value. To determine the amplitude of the current we use the formalism of complex numbers also used in Problem 40. The quantity on the left-hand side of equation (2.39) is $B_S A (\omega_0 - \omega) \cos\left[(\omega_0 - \omega)t + \theta_{S0} - \theta_{n0} - \frac{\pi}{2}\right]$. Its complex representative is therefore $\underline{\varepsilon}_i = B_S A (\omega_0 - \omega) \exp\left[i\left(\theta_{S0} - \theta_{n0} - \frac{\pi}{2}\right)\right]$. It then follows from equation (2.39) that

$$B_S A (\omega_0 - \omega) \exp\left[i\left(\theta_{S0} - \theta_{n0} - \frac{\pi}{2}\right)\right] = R\underline{I} + L(\omega_0 - \omega)i\underline{I},$$ (2.40)

where $\underline{I}$ is the complex representative of the current i_C, which leads to

$$\underline{I} = \frac{B_S A (\omega_0 - \omega)}{R + L(\omega_0 - \omega)i} \exp\left[i\left(\theta_{S0} - \theta_{n0} - \frac{\pi}{2}\right)\right].$$ (2.41)

The amplitude of the current is then

$$I_m = \frac{B_S A (\omega_0 - \omega)}{\sqrt{R^2 + L^2 (\omega_0 - \omega)^2}}.$$ (2.42)

(d) The torque of the force that acts on the rotor is

$$\vec{M} = \vec{p}_m \times \vec{B}_S,$$ (2.43)

where $\vec{p}_m = i_C A \vec{n}$ is the magnetic moment of the contour. This leads to

$$\vec{M} = i_C A B_S \sin[(\omega_0 - \omega)t + \theta_{S0} - \theta_{n0}] \vec{e}_z.$$ (2.44)

Using the solution of part (c), it follows that the dependence of current on time is given as

$$i_C(t) = I_m \cos[(\omega_0 - \omega)t + \varphi], \qquad (2.45)$$

where $\varphi = \theta_{S0} - \theta_{n0} - \frac{\pi}{2} - \varphi_0$ and φ_0 is the phase of the complex number $R + L(\omega_0 - \omega)i$. Using the trigonometric identity $\sin x \cos y = \frac{1}{2}[\sin(x-y) + \sin(x+y)]$, it follows from equations (2.44) and (2.45) that

$$M_z = \frac{1}{2}AI_mB_S\{\cos\varphi_0 + \sin[2(\omega_0 - \omega)t + \theta_{S0} - \theta_{n0} + \varphi]\}. \qquad (2.46)$$

Since the mean value of sinusoidal quantity during one period is zero, it follows that

$$M_{av} = \frac{1}{2}AI_mB_S\cos\varphi_0. \qquad (2.47)$$

Using $\cos\varphi_0 = \dfrac{R}{\sqrt{R^2 + L^2(\omega_0 - \omega)^2}}$ and equation (2.42) it follows that

$$M_{av} = \frac{1}{2}(B_SA)^2 \frac{R(\omega_0 - \omega)}{R^2 + L^2(\omega_0 - \omega)^2}, \qquad (2.48)$$

and finally

$$M_{av} = \frac{1}{2}(B_SA)^2 \frac{\omega_0}{R} \frac{s}{1 + \left(\frac{L\omega_0}{R}\right)^2 s^2}. \qquad (2.49)$$

(e) M_{av} is maximal when the function $f(s) = \dfrac{s}{1 + \left(\frac{L\omega_0}{R}\right)^2 s^2}$ reaches maximum. The first derivative of this function is

$$f'(s) = \frac{1 - \left(\frac{L\omega_0}{R}\right)^2 s^2}{\left[1 + \left(\frac{L\omega_0}{R}\right)^2 s^2\right]^2}. \qquad (2.50)$$

We conclude from the previous expression that $f'(s) = 0$ for $s = s_m = \frac{R}{L\omega_0}$, while $f'(s) > 0$ for $s < s_m$ and $f'(s) < 0$ for $s > s_m$. It then follows that the function $f(s)$ exhibits the maximum for $s = \frac{R}{L\omega_0}$.

(f) Newton's second law for rotation of the contour reads

$$J\frac{d\omega}{dt} = M_{av} - K\omega, \qquad (2.51)$$

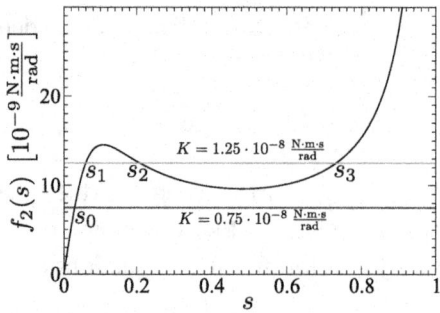

Figure 2.15 The graph of the dependence $f_2(s)$. Intersections of the graph with the line $f_2(s) = K$ for $K = 0.75 \cdot 10^{-8} \; \frac{\text{N·m·s}}{\text{rad}}$ and $K = 1.25 \cdot 10^{-8} \; \frac{\text{N·m·s}}{\text{rad}}$ are also shown in the figure

where J is the moment of inertia of the contour with respect to the rotation axis. In steady state we have $\frac{d\omega}{dt} = 0$, which leads to

$$K = \frac{1}{2} \frac{(B_S A)^2}{R} \frac{s}{(1-s)\left[1 + \left(\frac{L\omega_0}{R}\right)^2 s^2\right]}. \tag{2.52}$$

The graph of the function $f_2(s) = \frac{1}{2}\frac{(B_S A)^2}{R}\frac{s}{(1-s)\left[1+\left(\frac{L\omega_0}{R}\right)^2 s^2\right]}$ is shown in Figure 2.15. By solving the equation $K = f_2(s)$ using the graphical method, we find that it has one solution: $s_0 \approx 0.029$.

(g) By solving the equation $K = f_2(s)$ using the graphical method we find three solutions: $s_1 \approx 0.060$, $s_2 \approx 0.21$ and $s_3 \approx 0.73$.

(h) Newton's second law for rotation of the rotor reads

$$J\frac{d\omega}{dt} = M_{\text{av}}(\omega) - K\omega. \tag{2.53}$$

To investigate the stability of the system, we assume that the angular velocity was slightly changed in a certain moment of time from ω to $\omega + u$. The equation of motion then becomes:

$$J\frac{d(\omega + u)}{dt} = M_{\text{av}}(\omega + u) - K(\omega + u). \tag{2.54}$$

By subtracting equations (2.53) and (2.54) we obtain

$$J\frac{du}{dt} = M_{\text{av}}(\omega + u) - M_{\text{av}}(\omega) - Ku. \tag{2.55}$$

Since $M_{av}(\omega+u) - M_{av}(\omega) = \frac{dM_{av}}{d\omega} \cdot u$ holds for small u, it follows that

$$J\frac{du}{dt} = \left(\frac{dM_{av}}{d\omega} - K\right)u. \tag{2.56}$$

When

$$\frac{dM_{av}}{d\omega} - K < 0, \tag{2.57}$$

u will exponentially decrease in time, which means that the perturbation will vanish after some time and the system will return to equilibrium trajectory, which means it is stable. In the opposite case, u will exponentially increase in time and the system is unstable. By replacing equations (2.49) and (2.52) in equation (2.57) we obtain the stability condition

$$\left(\frac{L\omega_0}{R}\right)^2 (2s^3 - s^2) + 1 > 0. \tag{2.58}$$

This condition is satisfied for the solution s_0 in part (f), as well as for solutions s_1 and s_3 in part (g), while it is not satisfied for solution s_2 in part (g).

This problem was created by modifying a problem given at the World Physics Olympiad in 2013.

Problem 9   Internal Combustion Engines

What is horsepower?
Which engine is more efficient: Otto or Diesel?

Internal combustion engines are used in cars and other vehicles and in many machines. Their function is to transform the chemical energy of fuel into mechanical work. The scheme of the internal combustion engine is given in Figure 2.16. The main types of internal combustion engines are the Otto engine and the Diesel engine.

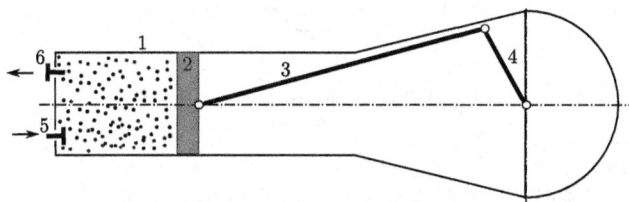

Figure 2.16 A scheme of the internal combustion engine: (1) cylinder, (2) piston, (3) connecting rod, (4) crankshaft, (5) inlet valve, (6) exhaust valve

The first modern internal combustion engine was constructed in 1877 by Nikolaus Otto. Its production started one year later in the USA. Different phases during the operation of a four-stroke Otto engine are shown in Figure 2.17.

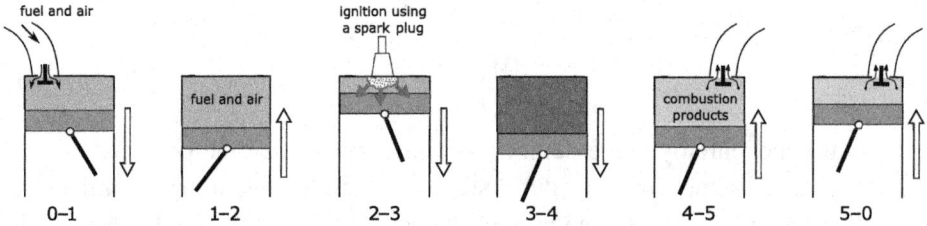

Figure 2.17 Phases during the operation of a four-stroke Otto engine

The process 0–1 corresponds to intake of the mixture of air and fuel from the carburetor to the cylinder of the engine. The mixture of air and fuel exhibits adiabatic compression in the process 1–2. The mixture is ignited by a spark plug and burns in an isohoric process 2–3. The combustion products expand adiabatically during the working stroke 3–4. The exhaust valve opens in the process 4–5 and combustion products flow out in an isohoric process. The remaining combustion products are removed from the cylinder in the process 5–0. The whole cycle can be represented using a p–V diagram as shown in Figure 2.18. The work gained in the intake stroke (process 0–1) is approximately equal to the work spent in the blowing-out stroke (process 5–0). For this reason, it is common to consider the Otto cycle without these two processes. Consider an ideal Otto cycle with air as the working fluid where the thermodynamic quantities are the same in points 1 and 5.

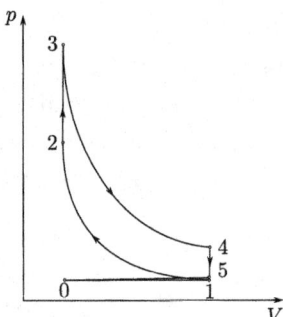

Figure 2.18 The diagram of the Otto cycle

(a) Plot the $p - V$ and $T - S$ diagram of an ideal Otto cycle.
(b) Express the efficiency of this cycle in terms of the compression index $b = V_{max}/V_{min}$ and the adiabatic index κ.

(c) Calculate the useful work and power in an ideal Otto cycle that corresponds to a one-cylinder four-stroke Otto engine. The diameter of the cylinder is $D = 86.0\,\text{mm}$, the stroke of the cylinder is $s = 100\,\text{mm}$, and the frequency of the rotation of the crankshaft is $n = 1,800\,\text{turns/min}$. The initial pressure and temperature in the cycle are $p_1 = 1.00\,\text{bar}$ and $T_1 = 47.0\,°\text{C}$, the maximal temperature is $T_{\text{max}} = 1,800\,°\text{C}$, and the compression index is $b = 6.20$. Assume that the air is an ideal gas of molar mass $M = 28.9\,\text{g/mol}$ and specific heat capacity $c_v = 720\,\text{J/(kg·K)}$. The gas constant is $R = 8.314\,\text{J/(mol·K)}$, $0\,°\text{C} = 273.15\,\text{K}$.

Rudolf Diesel was working on the development of a four-stroke engine the cycle of which would be as close as possible to the Carnot cycle. In 1890 he got an idea for the diesel engine as we know it today. Two years later he patented this engine in Germany.

(d) With the support of big companies, Diesel produced a series of more successful engines. This culminated in 1897 when he made a one cylinder four-stroke engine of 25 horsepower. What was the power of this engine expressed in Watts? The horsepower is an old unit for power that is occasionally used nowadays by some engineers. One horsepower is the power produced by an average horse, which lifts the weight of mass $m = 75\,\text{kg}$ (Figure 2.19), in $t = 1\,\text{s}$ by $h = 1\,\text{m}$. Gravitational acceleration is $g = 9.81\,\text{m/s}^2$.

Figure 2.19 The horse that lifts the weight

The phases of operation of the diesel engine are shown in Figure 2.20. The difference with respect to the Otto engine is that there is only the intake of air in

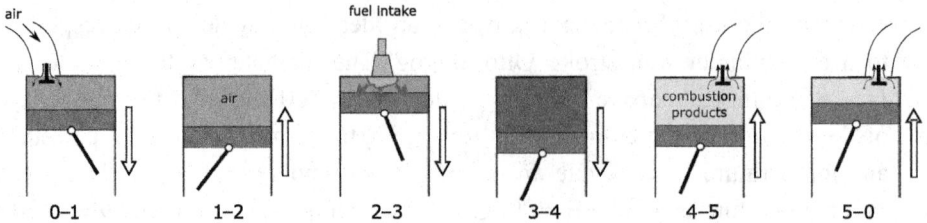

Figure 2.20 The phases of operation of a four-stroke Diesel engine

the process 0–1, while a specific device is used to inject a certain amount of fuel – oil in the process 2–3 which is now isobaric. Due to high injection pressure, the oil scatters out and immediately mixes with hot air in the cylinder. The contact of drops of fuel with hot air leads to partial evaporation of the fuel followed by its self-ignition and burning. Other processes are identical to those that occur in the Otto engine.

(e) Plot the $p-V$ diagram of the Diesel cycle, assuming that points 1 and 5 have similar thermodynamic quantities. Plot also the $p-V$ and $T-S$ diagram of an ideal Diesel cycle.

(f) Determine the efficiency of the ideal Diesel cycle in terms of the compression index $b = V_{max}/V_{min} = V_1/V_2$, the degree of charging $a = V_3/V_2$, and the adiabatic index κ of the ideal working fluid.

The combined cycle is the basis for the operation of modern diesel engines with injection of fuel in the antechamber using a high-pressure pump. The antechamber is connected to the main combustion chamber by several channels. This cycle is a combination of the Otto and the Diesel cycles: heat is initially brought in the isohoric process (process 2–3) which corresponds to immediate combustion in the antechamber, while, when a certain pressure is reached, heat is brought in the isobaric process (process 3′–3), which corresponds to combustion in the main combustion chamber.

(g) Plot the $p-V$ diagram of the combined cycle. Determine the efficiency of this cycle in terms of the compression index $b = V_{max}/V_{min}$, the degree of charging $a = V_3/V_{3'}$, the pressure ratio $c = p_{3'}/p_2$, and the adiabatic index κ of the ideal working fluid.

Solution of Problem 9

(a) $p-V$ and $T-S$ diagrams of the ideal Otto cycle are shown in Figure 2.21.

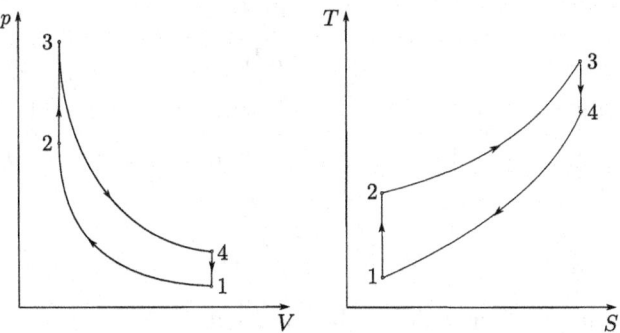

Figure 2.21 $p-V$ and $T-S$ diagrams of the ideal Otto cycle

(b) The efficiency is

$$\eta = 1 - \frac{|Q_{\text{pred}}|}{Q_{\text{prim}}}, \tag{2.59}$$

where Q_{prim} is the total heat that the gas received from the surroundings, while Q_{pred} is the heat that the gas gave to the surroundings during the cycle. In the ideal Otto cycle, the gas receives heat in process 2–3, it gives the heat away in process 4–1, and there is no exchange of heat in the adiabatic processes 1–2 and 3–4, which implies

$$\eta = 1 - \frac{|Q_{41}|}{Q_{23}} = 1 - \frac{mc_{\text{v}}(T_4 - T_1)}{mc_{\text{v}}(T_3 - T_2)}. \tag{2.60}$$

Points 1(3) and 2(4) belong to the same adiabat, which implies

$$T_{1(3)}V_{1(3)}^{\kappa-1} = T_{2(4)}V_{2(4)}^{\kappa-1}, \tag{2.61}$$

and leads to $T_1/T_2 = T_4/T_3 = b^{1-\kappa}$. The efficiency of the ideal Otto cycle is then

$$\eta = 1 - \frac{T_1(T_4/T_1 - 1)}{T_2(T_3/T_2 - 1)} = 1 - \frac{T_1}{T_2} = 1 - b^{1-\kappa}. \tag{2.62}$$

The increase of the compression index leads to an increase of efficiency. However, the pressure and the temperature of the mixture increase at the same time, which can lead to a negative consequence – self-ignition of the mixture. Such a process is accompanied by a detonation and can cause mechanical damage of the engine. For this reason, the compression index of these engines must be smaller than 10–12, depending on the quality of the fuel.

(c) In an ideal Otto cycle, the difference of the heat received and the heat given is equal to the useful work $A = Q_{\text{prim}} - Q_{\text{pred}} = mc_{\text{v}}(T_3 - T_2 - T_4 + T_1)$. The mass of the working fluid is obtained from ideal gas law for state 1

$$m_1 = \frac{Mp_1V_1}{RT_1}, \tag{2.63}$$

while the adiabatic index is $\kappa = \frac{Mc_v+R}{Mc_v}$. The difference of the maximal and minimal volume is $V_1 - V_2 = \frac{\pi D^2}{4}s$, while their ratio is $V_1/V_2 = b$, which leads to

$$V_1 = \frac{\pi b D^2 s}{4(b-1)}. \tag{2.64}$$

The temperatures T_2 and T_4 can be obtained from equation (2.61), and we finally obtain that the useful work is

$$A = \frac{\pi M p_1 b D^2 s c_v}{4(b-1)R}\left[\frac{T_3}{T_1}(1-b^{1-\kappa}) - b^{\kappa-1}+1\right] = 395\,\text{J}. \tag{2.65}$$

In four-stroke engines one period of the crankshaft corresponds to processes 0–1, 1–2, and 2–3, while its other period corresponds to processes 3–4, 4–5, and 5–0. For this reason, the period of the cycle is twice the period of the crankshaft

$$t_1 = 2T = \frac{2}{n}, \tag{2.66}$$

which implies that the power in question is

$$P = \frac{A}{t_1} = \frac{\pi M p_1 b D^2 s c_v n}{8(b-1)R}\left[\frac{T_3}{T_1}(1-b^{1-\kappa}) - b^{\kappa-1}+1\right] = 5.93\,\text{kW}. \tag{2.67}$$

(d) The power of the horse is $P = mgh/t$, where g is gravitational acceleration. The required ratio is $1\,\text{hp} = 736\,\text{W}$. Therefore, the most powerful diesel engine had a power of $P_D = 25 \cdot 736\,\text{W} = 18.4\,\text{kW}$.

(e) The $p-V$ diagram of the Diesel cycle as well as the $p-V$ and $T-S$ diagram of the ideal Diesel cycle are shown in Figure 2.22.

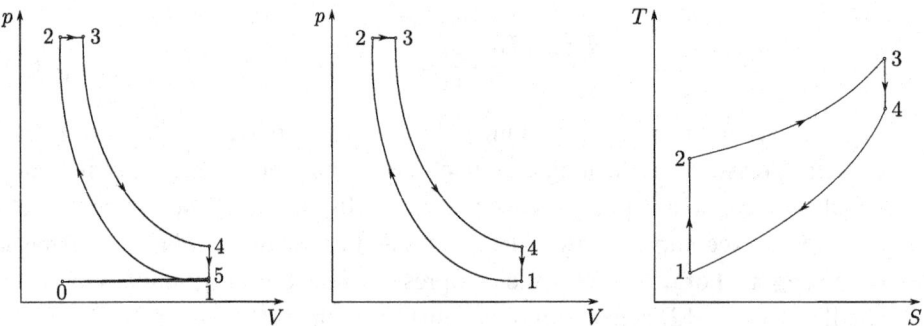

Figure 2.22 $p-V$ diagram of the Diesel cycle; $p-V$ and $T-S$ diagrams of the ideal Diesel cycle

(f) Using the analogy with ideal Otto cycle from part (b) of the problem, we find that the efficiency is

$$\eta = 1 - \frac{|Q_{41}|}{Q_{23}} = 1 - \frac{mc_v(T_4 - T_1)}{mc_p(T_3 - T_2)} = 1 - \frac{1}{\kappa} \frac{(T_4 - T_1)}{(T_3 - T_2)}. \tag{2.68}$$

Points 1 and 2 as well as points 3 and 4 are on the same adiabat and consequently equation (2.61) holds while points 2 and 3 are on the same isobar and therefore

$$\frac{T_2}{T_3} = \frac{V_2}{V_3} = \frac{1}{a}. \tag{2.69}$$

We can express all temperature in terms of T_1 using equations (2.61) and (2.69). We then obtain the efficiency

$$\eta = 1 - \frac{1}{\kappa} \frac{(a^\kappa T_1 - T_1)}{(ab^{\kappa-1}T_1 - b^{\kappa-1}T_1)} = 1 - \frac{a^\kappa - 1}{\kappa b^{\kappa-1}(a-1)}. \tag{2.70}$$

(g) The $p - V$ diagram of the combined cycle is shown in Figure 2.23. The efficiency of this cycle is

$$\eta = 1 - \frac{|Q_{41}|}{Q_{23}} = 1 - \frac{mc_v(T_4 - T_1)}{mc_v(T_{3'} - T_2) + mc_p(T_3 - T_{3'})}$$

$$= 1 - \frac{(T_4 - T_1)}{(T_{3'} - T_2) + \kappa(T_3 - T_{3'})}. \tag{2.71}$$

We want to express all temperatures in terms of T_1. Points 1 and 2 as well as points 3 and 4 are on the same adiabat and equation (2.61) holds. Points 2 and $3'$ are on the same isohor, while $3'$ and 3 are on the same isobar, which leads to

$$\frac{T_{3'}}{T_2} = \frac{p_{3'}}{p_2} = c \text{ and } \frac{T_3}{T_{3'}} = \frac{V_3}{V_{3'}} = a. \tag{2.72}$$

We find from equations (2.61) and (2.72) that $T_2 = b^{\kappa-1}T_1$, $T_{3'} = cb^{\kappa-1}T_1$ and $T_3 = acb^{\kappa-1}T_1$, while T_4 satisfies

$$T_4 = T_3 \left(\frac{V_3}{V_1}\right)^{\kappa-1} = cb^{\kappa-1}T_1 \left(\frac{V_3 V_{3'}}{V_{3'} V_1}\right)^{\kappa-1} = ca^\kappa T_1. \tag{2.73}$$

We finally obtain the efficiency

$$\eta = 1 - \frac{ca^\kappa - 1}{b^{\kappa-1}[c - 1 + \kappa c(a-1)]}. \tag{2.74}$$

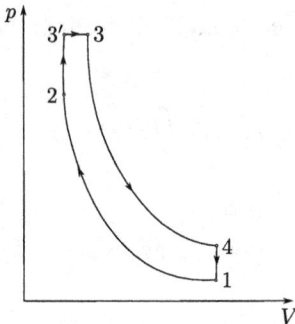

Figure 2.23   The $p-V$ diagram of the combined cycle

We note that for $c=1$ the efficiency of the combined cycle reduces to the efficiency of the Diesel cycle, while for $a=1$ it reduces to the efficiency of the Otto cycle. This analysis is useful as a check of the final result.

Problem 10   Air Conditioner

What is going on inside an air conditioner?
To what temperature can it cool a room?

Air conditioners are used to cool rooms on hot summer days. In this problem we investigate in detail the operating principles of an air conditioner. An air conditioner consists of a unit that is inside a room and a unit that is outside. The working fluid of the air conditioner that we consider is 1,1,1,2-tetrafluoroethane. Its temperature of evaporation at the pressure of $p_1=133\,\text{kPa}$ is $t_1=-20.0\,°\text{C}$, while it is equal to $t_3=40.0\,°\text{C}$ at the pressure of $p_3=1.00\,\text{MPa}$. The working fluid in the unit inside the room is at a temperature t_1 and consequently receives heat from the room, while the working fluid of the unit outside is at a temperature t_3 and therefore it gives heat to the environment. This way the working fluid effectively transfers heat from the room to the environment.

The scheme of the main components of an air-conditioner is given in Figure 2.24. The main component of the unit inside the room is the evaporator. In the stationary state of an operating air conditioner the fluid in the evaporator is in the state that is a mixture of the gas and the liquid at the temperature t_1 and the pressure p_1. At the exit of evaporator, the fluid is a gas at the temperature t_1 and the pressure p_1 (state 1). The fluid is taken from the evaporator to the compressor that is outside the room. The fluid is then compressed so that it is in a gaseous state at the pressure $p_2=p_3$ and the temperature $t_2=70.0\,°\text{C}$ (state 2) at the exit of the compressor. The fluid is taken from the compressor to the condenser where it loses heat (gives it to the

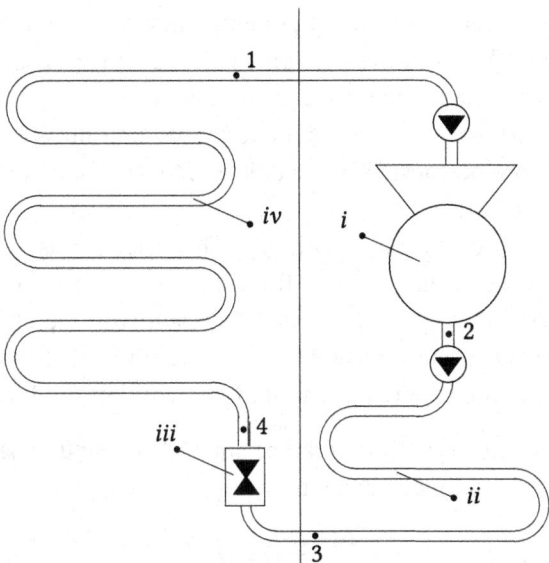

Figure 2.24 The scheme of the main components of an air conditioner: (i) compressor, (ii) condenser, (iii) throttle, (iv) evaporator. The components on the left comprise the inside unit while the components on the right belong to the outside unit

environment) at a constant pressure p_2. The fluid is in a liquid state of temperature t_3 and pressure p_3 at the exit of the condenser (state 3). Next the fluid is taken back to the evaporator through the throttle. This process is adiabatic and the temperature and pressure are reduced to $t_4 = t_1$ and $p_4 = p_1$ (state 4) where the fluid is in a state that is a mixture of the liquid and the gas. The fluid evaporates again in the evaporator and it is again in state 1 at the exit of evaporator. Subsequently the whole cycle repeats.

The specific heat of vaporization of the working fluid is $h_1 = 220 \frac{\text{kJ}}{\text{kg}}$ at the pressure p_1 and $h_3 = 160 \frac{\text{kJ}}{\text{kg}}$ at the pressure p_3. Assume that the working fluid in the gaseous state is an ideal gas whose adiabatic index is $\gamma = 1.10$. The molar mass of the working fluid is $M = 102 \frac{\text{g}}{\text{mol}}$. The gas constant is $R = 8.31 \frac{\text{J}}{\text{mol·K}}$, while $0\,°\text{C} = 273.15\,\text{K}$. Assume that the density of fluid in a gaseous state is much smaller than when the fluid is in a liquid state.

(a) Calculate the work a_{12} performed on the unit mass of the fluid in the compressor during fluid compression. Calculate also the work a_{23} performed on the unit mass of the fluid during its condensation in the condenser. Assume that the dependence of the pressure on the gas volume is linear during the compression in the compressor.

(b) Calculate the difference of entropy per unit mass of the fluid $s_1 - s_4$ in states 1 and 4, as well as the work a_{41} performed on the unit mass of the fluid during the transition from state 4 to state 1 in the evaporator.

(c) Determine the difference of internal energy per unit mass $u_3 - u_4$ in states 3 and 4 and the work performed on the unit mass of the fluid during the transition from state 3 to state 4.

(d) Find the efficiency of this air conditioner. The efficiency is defined as the ratio of the amount of heat that is taken from the room in one cycle and the work performed during that cycle. Compare the result with the efficiency of an air conditioner that utilizes the inverse Carnot cycle with the temperatures equal to the largest and smallest temperature of the working fluid during the cycle.

The law of heat conduction states that the amount of heat that passes through the walls of the room in unit time is equal to

$$\frac{\Delta Q}{\Delta t} = \alpha S \Delta T,$$

where α is the heat conduction coefficient that depends on the type of the material of the walls, S is the area of the outside walls of the room, and ΔT is the difference of the temperatures of the environment and the room.

The air conditioner considered in this problem cools a room with an outside wall area of $S = 50.0\,\text{m}^2$. The coefficient of heat conduction through the walls is $\alpha = 2.20\,\frac{\text{W}}{\text{m}^2\text{K}}$. The power of the air conditioner is $P = 1.00\,\text{kW}$.

(e) What is the minimal possible temperature of the room cooled by this air conditioner if the outside temperature is $t_s = 30.0°\text{C}$?

(f) What is the fraction of time that this air conditioner should be operating to cool the room to the temperature of $t_u = 25.0°\text{C}$?

Solution of Problem 10

(a) The dependence of the pressure on the volume is linear in the process 1–2 and the work performed on the fluid is equal to the negative value of the area under the curve that represents this dependence. The required work a_{12} is equal to the volume of the trapezoid with a height $v_1 - v_2$ and lengths of the bases p_1 and p_2. Consequently $a_{12} = \frac{1}{2}(p_1 + p_2)(v_1 - v_2)$, where v_1 and v_2 are the volumes of the unit mass of fluid in states 1 and 2. The equation of state for ideal gas yields $p_2 v_2 = RT_2/M$ and $p_1 v_1 = RT_1/M$. From previous equations we obtain

$$a_{12} = \frac{R}{2M}\left[T_1\left(1 + \frac{p_2}{p_1}\right) - T_2\left(1 + \frac{p_1}{p_2}\right)\right] = 72.0\,\frac{\text{kJ}}{\text{kg}}. \qquad (2.75)$$

In process 2–3, the gas first cools under the constant pressure p_3 and reaches the temperature t_3 (state $2'$). Then it condenses at a constant temperature and pressure until it reaches the liquid state (state 3). The work performed on the unit mass of gas in process 2–$2'$ is $a_{22'} = p_3 (v_2 - v_{2'})$. Using equations of state $p_3 v_2 = RT_2/M$ and $p_3 v_{2'} = RT_3/M$ we obtain $a_{22'} = \frac{R}{M}(T_2 - T_3) = 2.44 \frac{\text{kJ}}{\text{kg}}$. Next we have $a_{2'3} = p_3 (v_{2'} - v_3)$, where $v_3 \ll v_{2'}$ because the fluid is in the liquid state in state 3. Using the equation of state $p_3 v_{2'} = RT_3/M$, we find $a_{2'3} = RT_3/M = 25.5 \frac{\text{kJ}}{\text{kg}}$. We finally obtain $a_{23} = a_{22'} + a_{2'3} = 28.0 \frac{\text{kJ}}{\text{kg}}$.

(b) Since $s_1 - s_4 = (s_1 - s_{2'}) + (s_{2'} - s_3) + (s_3 - s_4)$, the difference required can be determined by finding each of the terms in brackets in the previous expression. Since process 3–4 is adiabatic, it follows that $s_3 - s_4 = 0$. Next, we have $s_{2'} - s_3 = h_3/T_3$ because process $2'$–3 is isothermal. To find the term $s_1 - s_{2'}$, we make use of the fact that entropy is the function of the state of the system. Consequently this term can be determined by finding it for any process that starts in state 1 and ends in state $2'$. Since it is easiest to calculate the change of entropy for isothermal and adiabatic processes, we construct the process in which the gas is in state 1 at the beginning, follows an isothermal process and reaches state x, and finally follows an adiabatic process and reaches state $2'$. The equation for adiabatic process reads $p_3 v_{2'}^{\gamma} = p_x v_x^{\gamma}$. The equations of state for states 1, $2'$, and x are $p_1 v_1 = RT_1/M$, $p_3 v_{2'} = RT_3/M$ and $p_x v_x = RT_1/M$. The change of entropy per unit mass in an isothermal process is in the case of ideal gas equal to $s_x - s_1 = \frac{R}{M}\ln(v_x/v_1)$, while $s_{2'} - s_x = 0$ for an adiabatic process. Previous equations yield $s_{2'} - s_1 = \frac{R}{M}\ln\left[\frac{p_1}{p_3}\left(\frac{T_3}{T_1}\right)^{\frac{\gamma}{\gamma-1}}\right]$. We then finally obtain

$$s_1 - s_4 = h_3/T_3 - \frac{R}{M}\ln\left[\frac{p_1}{p_3}\left(\frac{T_3}{T_1}\right)^{\frac{\gamma}{\gamma-1}}\right] = 485 \frac{\text{J}}{\text{kg}\cdot\text{K}}. \tag{2.76}$$

Next we have $a_{41} = p_1 (v_4 - v_1)$. The difference of entropy per unit mass for liquid (s_L) and gaseous state (s_1) at a temperature t_1 and pressure p_1 is $s_1 - s_L = h_1/T_1 = 869 \frac{\text{J}}{\text{kg}\cdot\text{K}}$. Since $s_1 - s_4 < s_1 - s_L$, we conclude that the fluid is a mixture of the liquid and the gas in state 4, where the mass fraction of the gaseous state is $r = 1 - \frac{s_1 - s_4}{s_1 - s_L} = 44.2\%$. By neglecting the volume of the liquid it follows that $v_4 = rv_1$, and consequently $a_{41} = p_1 v_1 (r - 1)$. Using the equation of state $p_1 v_1 = RT_1/M$, it follows that $a_{41} = \frac{RT_1}{M}(r - 1) = -11.5 \frac{\text{kJ}}{\text{kg}}$.

(c) The difference $u_3 - u_4$ can be found from $u_3 - u_4 = (u_3 - u_{2'}) + (u_{2'} - u_1) + (u_1 - u_4)$. The first law of thermodynamics for process $2'$–3 reads $u_3 - u_{2'} = a_{2'3} - h_3 = -134.5 \frac{\text{kJ}}{\text{kg}}$. The internal energy per unit mass of ideal gas is given as $u = C_v T/M$, where C_v is the molar heat capacity at a constant volume $C_v = R/(\gamma - 1)$. Consequently $u_{2'} - u_1 = \frac{R}{M(\gamma-1)}(T_3 - T_1) = 48.9 \frac{\text{kJ}}{\text{kg}}$. The first law of thermodynamics for process 4–1 gives $u_4 - u_1 = -T_1(s_1 - s_4) - a_{41} = -111.2 \frac{kJ}{kg}$. By taking into account all of these contributions, it follows that $u_3 - u_4 = 25.6 \frac{kJ}{kg}$. The first law of thermodynamics for adiabatic process 3–4 gives $a_{34} = -(u_3 - u_4) = -25.6 \frac{kJ}{kg}$.

(d) The efficiency of this air conditioner is $\eta = \frac{T_1(s_1-s_4)}{a_{41}+a_{12}+a_{23}+a_{34}} = 1.95$. The efficiency of the device that uses the inverse Carnot cycle is larger, and it is equal to $\eta_c = \frac{T_1}{T_2-T_1} = 2.81$.

(e) The lowest temperature of the room is achieved when the air conditioner operates all the time on full power. In steady state, the amount of heat that the air conditioner gives to the environment is equal to the amount of heat that enters the room through the walls. This condition gives $\eta P = \alpha S(t_s - t_{min})$, and consequently $t_{min} = t_s - \frac{\eta P}{\alpha S} = 12.3°C$.

(f) When the air conditioner operates only part of the time, the equality from the previous part is modified to $\beta \eta P = \alpha S(t_s - t_u)$, where $0\% \leq \beta \leq 100\%$ is the fraction of time when the air conditioner operates. Consequently we obtain $\beta = \frac{\alpha S(t_s - t_u)}{\eta P} = 28.2\%$.

We refer the reader interested in more details about air conditioners to references [3] and [26]. The authors presented this problem at the Serbian Physics Olympiad for high school students in 2017.

Problem 11   Motion of a Car

What is the power required for a car to travel at $100 \frac{km}{h}$?
What torque on the wheels is required?

The motion of a car is considered in this problem. In particular, we determine the power needed for a car to travel at a speed of $100 \frac{km}{h}$ and the torque that needs to act on wheels of the car in that case.

(a) Make an order-of-magnitude estimate of the useful power that has to be produced by the engine of a car for the car to travel at a speed of $100 \frac{km}{h}$. In this part of the problem do not use the information given in the rest of the problem.

A four-wheeled car whose cross-section is shown in Figure 2.25 is moving on a flat road. The total mass of the car is $m = 1,580\,kg$, the mass of each wheel is $m_t = 20.0\,kg$, and the mass of the axles that connect the wheels can be neglected. Assume that the wheels are in the shape of homogeneous thin cylinders of radius $R = 18.0\,cm$. This car is a front-wheel drive car, which means that the engine provides a torque of magnitude M_0 that acts on the front wheels and enables their

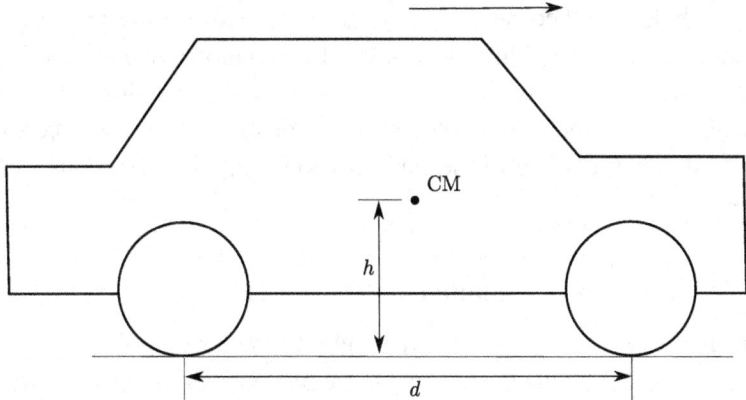

Figure 2.25 Cross-section of a car. CM denotes the position of the center of mass

rotation around the axle that connects them. The center of mass of the car is at a height of $h = 1.00$ m above the ground (see Figure 2.25) and it is located in the vertical plane that halves the line that connects the front and the rear right (left) wheel. The distance between the front and the rear pair of wheels is $d = 3.00$ m (Figure 2.25). There is no sliding between the wheels and the ground. Gravitational acceleration is $g = 9.81\,\frac{m}{s^2}$.

We consider first the motion of a car that was initially at rest. Therefore, its speed is small and air resistance can be neglected. The magnitude of the torque that acts on the front wheels is $M_0 = 270\,\mathrm{N \cdot m}$.

(b) Calculate the magnitude of the acceleration of the car.
(c) Determine the magnitude and the direction of the friction force on each of the wheels.
(d) Calculate the reaction force between each of the wheels and the ground.
(e) For which values of the friction coefficient between the wheels and the ground is there no sliding?
(f) Determine the useful work per unit of distance traveled that should be produced by the engine of the car to enable this motion.
(g) Which part of useful work is spent on the kinetic energy of translation of the car and which part is spent on the kinetic energy of rotation of the wheels?

We next consider a car that is traveling at a speed $v = 100\,\frac{km}{h}$. The magnitude of the air drag force that acts on this car is $F_{ot} = cv^2$, where $c = 0.900\,\frac{kg}{m}$.

(h) Calculate the magnitude of the torque M_0 that enables such motion of the car.
(i) Calculate the useful work per unit of distance traveled and the useful power that should be produced by the engine of the car to enable this motion. If 1.00 l of gasoline is consumed to provide useful work of 10.0 MJ, how many liters of gasoline will be consumed by this car when it travels a distance of 100 km?

(j) How much does fuel consumption increase (decrease) if the car is traveling at the same speed on the uphill (downhill) of inclination angle of $\varphi = 2.00°$?

(k) A car is traveling on a horizontal road. Its reservoir contains only $V = 1.00\,\mathrm{l}$ of gasoline, while the closest gas station is at a distance of $s = 20.0\,\mathrm{km}$. What strategy should the driver of the car apply to reach the gas station?

Solution of Problem 11

(a) The engine needs to produce a force equal to the air drag force at a speed of $v = 100\,\frac{\mathrm{km}}{\mathrm{h}}$. We estimate the air drag force by considering the collisions of the particles of air with the car. The particles that collide with the car in time Δt are all air particles from the volume $\Delta V = Sv\Delta t$, where S is the area of the cross-section of the car perpendicular to the direction of motion. The mass of these particles is $\Delta m = \rho \Delta V$, where ρ is the density of air. These particles transfer the momentum $\Delta p \sim \Delta mv$ to the car. Therefore, the force between the particles and the car is $F = \frac{\Delta p}{\Delta t}$. The useful power that has to be produced by the engine is $P = Fv$. We find from previous equations that $P \sim \rho Sv^3$. We estimate the numerical values of the quantities from the previous equations as $\rho \sim 1\,\frac{\mathrm{kg}}{\mathrm{m}^3}$, $S \sim 2\,\mathrm{m}^2$ and $v = 100\,\frac{\mathrm{km}}{\mathrm{h}}$, which leads to $P \sim 40\,\mathrm{kW}$.

(b) Newton's second law for rotation of the front wheels yields:

$$I\alpha = M_0 - 2FR, \qquad\qquad (2.77)$$

where $I = 2 \cdot \frac{1}{2}m_t R^2$ is the moment of inertia of the front pair of wheels with respect to the rotation axis, α is the magnitude of angular acceleration of the wheels, and F is the projection of the friction force between the front wheel and the ground on the direction of motion. The direction of this force that is assumed is the direction of motion of the car, and it is shown in Figure 2.26.

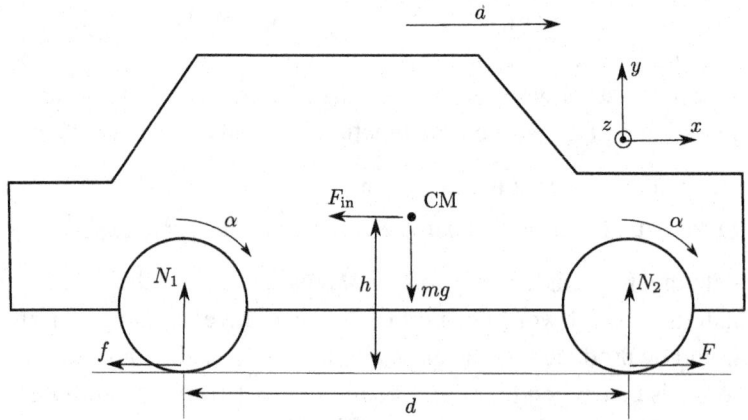

Figure 2.26 Cross-section of the car and the forces that act on it

Newton's second law for the rear wheels gives

$$I\alpha = 2fR, \tag{2.78}$$

where f is the projection of friction force between the rear wheel and the ground on the direction of motion. The direction of this force that is assumed is opposite to the direction of motion of the car (Figure 2.26).

Newton's second law for translation of the car reads

$$ma = 2F - 2f, \tag{2.79}$$

where a is the magnitude of acceleration of the car. Since there is no sliding between the wheels and the ground, it follows that

$$a = R\alpha. \tag{2.80}$$

Previous equations imply

$$a = \frac{M_0}{R(m + 2m_t)} = 0.926 \, \frac{\text{m}}{\text{s}^2}. \tag{2.81}$$

(c) It follows from equations (2.77)–(2.80) that

$$F = \frac{M_0}{2R} \frac{m + m_t}{m + 2m_t} = 741 \, \text{N} \tag{2.82}$$

and

$$f = \frac{M_0}{2R} \frac{m_t}{m + 2m_t} = 9.26 \, \text{N}. \tag{2.83}$$

Since the values obtained are positive, it follows that the directions of friction forces coincide with the directions assumed that are presented in Figure 2.26.

(d) The forces that act on the car in non-inertial reference frame of the car are the gravity force directed downward, the reaction force of the ground on each of the wheels, the friction force between the wheels and the ground, and the inertial force the magnitude of which is $F_{\text{in}} = ma$ and it is directed opposite to the direction of motion of the car; see Figure 2.26. Since the car is at rest in this reference frame, the sum of these forces and the sum of torques of these forces must be equal to zero. From the condition that the projection of the sum of all forces on vertical direction is zero, we obtain:

$$mg = 2N_1 + 2N_2, \tag{2.84}$$

where N_1 is the magnitude of the reaction force acting on the rear wheel, while N_2 is the magnitude of the reaction force acting on the front wheel. From the condition that the z-component of the sum of torques with respect to the point of contact of the front wheel and the ground is zero, we find

$$2N_1 d - mg\frac{d}{2} - mah = 0. \tag{2.85}$$

From equations (2.84) and (2.85) it follows that

$$N_1 = \frac{mg}{4} + \frac{mah}{2d} = 4.12\,\text{kN} \tag{2.86}$$

and

$$N_2 = \frac{mg}{4} - \frac{mah}{2d} = 3.63\,\text{kN}. \tag{2.87}$$

(e) To prevent sliding, the conditions $f \le \mu N_1$ and $F \le \mu N_2$ need to be satisfied, where μ is the friction coefficient. These conditions imply

$$\mu \ge \frac{F}{N_2} = 0.204. \tag{2.88}$$

(f) The useful work of the engine is $A_k = M_0 \theta$, where θ is the angle of rotation of the front wheels. Since there is no sliding between the wheels and the ground, the distance traveled by the car is $s = R\theta$. The useful work per unit of distance traveled is then $F_k = \frac{A_k}{s} = \frac{M_0}{R} = 1.50\,\frac{\text{MJ}}{\text{km}}$.

(g) The speed of the car is $v = at$ after time t, while the magnitude of the angular velocity is $\omega = \alpha t$, and the distance traveled is $s = \frac{1}{2}at^2$. The useful work per unit of distance traveled that is spent on the kinetic energy of translation is $F_{\text{tr}} = \frac{T_{\text{tr}}}{s}$, where $T_{\text{tr}} = \frac{1}{2}mv^2$ is the kinetic energy of translation of the car. Previous equations imply that $F_{\text{tr}} = ma$. The useful work per unit of distance traveled that is spent on the kinetic energy of rotation of the wheels is $F_{\text{rot}} = \frac{T_{\text{rot}}}{s}$, where $T_{\text{rot}} = 2 \cdot \frac{1}{2}I\omega^2$ is the kinetic energy of rotation of the wheels, where factor 2 is present because both front and rear wheels rotate. From previous equations and equation (2.80) we find $F_{\text{rot}} = 2m_t a$. The fraction of useful work that is spent on the kinetic energy of translation is therefore $r_{\text{tr}} = \frac{F_{\text{tr}}}{F_{\text{tr}} + F_{\text{rot}}}$, which leads to $r_{\text{tr}} = \frac{m}{m + 2m_t} = 97.5\%$. The fraction of useful work that is spent on the kinetic energy of rotation of the wheels is then $r_{\text{rot}} = \frac{F_{\text{rot}}}{F_{\text{tr}} + F_{\text{rot}}}$ – that is, $r_{\text{rot}} = \frac{2m_t}{m + 2m_t} = 2.5\%$.

(h) Since the velocity of the car is constant, its acceleration and angular acceleration are zero. Equation (2.77) then takes the form

$$0 = M_0 - 2FR, \tag{2.89}$$

equation (2.78) reduces to

$$f = 0. \tag{2.90}$$

Newton's second law for motion of the car reads:

$$0 = 2F - 2f - cv^2. \tag{2.91}$$

We obtain from previous equations that

$$M_0 = cRv^2 = 125\,\text{N} \cdot \text{m}. \tag{2.92}$$

(i) From the solution of part (f), it follows that the useful work per unit of distance traveled is $F_k = \frac{M_0}{R}$, which leads to $F_k = cv^2 = 694\frac{\text{kJ}}{\text{km}}$. The useful work that is spent on the distance of 100 km is then $A_k = 69.4\,\text{MJ}$. Since 1.00 l of gasoline is consumed for production of 10.0 MJ, it follows that 6.94 l are needed to produce the work A_k. The useful power is $P = F_k v = cv^3 = 19.3\,\text{kW}$. We see that the order of magnitude of this result agrees with the estimate from part (a).

(j) Newton's second law for motion of the car at a constant speed uphill (downhill) reads:

$$0 = 2F - 2f \mp mg\sin\varphi - cv^2, \tag{2.93}$$

where the minus sign refers to the uphill and the plus sign to the downhill. Using equations (2.89) and (2.90) it follows that

$$M_0 = \left(cv^2 \pm mg\sin\varphi\right)R. \tag{2.94}$$

It follows from previous parts of the problem that the consumption of fuel is directly proportional to the torque. For this reason, consumption of fuel increases

$$r_u = \frac{cv^2 + mg\sin\varphi}{cv^2} = 1.78 \tag{2.95}$$

times uphill, while it decreases

$$r_n = \frac{cv^2}{cv^2 - mg\sin\varphi} = 4.52 \tag{2.96}$$

times downhill.

(k) Since only $V = 1.00\,\text{l}$ of gasoline remains in the reservoir, the useful work that can be produced by the engine is 10.0 MJ. It follows that the car must move in such a way that the useful work per unit of distance traveled is smaller than $F_k = \frac{10.0\,\text{MJ}}{20.0\,\text{km}} = 0.500\frac{\text{MJ}}{\text{km}}$. Since it was shown in part (i) that $F_k = cv^2$, we conclude that the car should move at a speed that is smaller than $v_{\max} = \sqrt{\frac{F_k}{c}} = 84.9\,\frac{\text{km}}{\text{h}}$.

We refer the interested reader to reference [14].

Problem 12   Airplane and Helicopter

What is the origin of the lift force in an airplane and a helicopter?

The lift force acting on an airplane is caused by the interaction between the wings and the air that flows around the airplane. We assume that air is an ideally incompressible fluid whose density is ρ and that it is only the air inside the region whose cross-section is the ellipse shown in Figure 2.27 that interacts with the wings of the plane. The length of the major axis of the ellipse is equal to the wingspan $2b$, while the length of the minor semi-axis is h. Assume that h is a harmonic average of the quantities b and c (shown in Figure 2.27):

$$\frac{1}{h} = \frac{1}{2}\left(\frac{1}{b} + \frac{1}{c}\right). \tag{2.97}$$

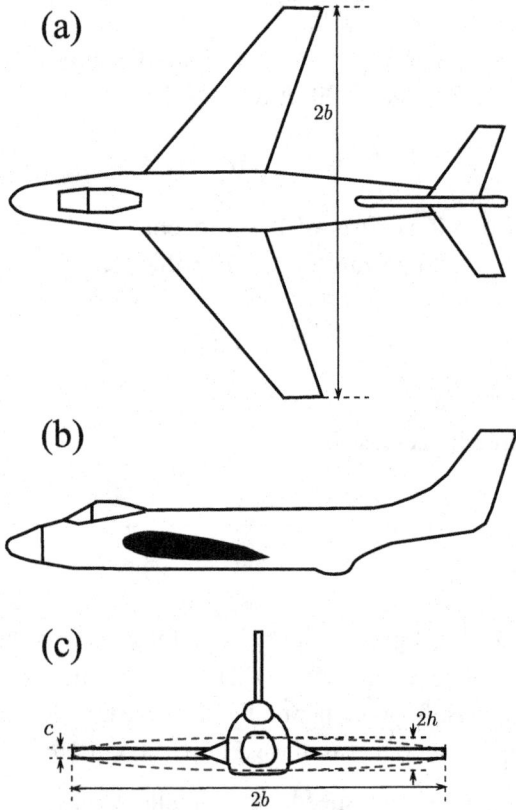

Figure 2.27 Scheme of an airplane: (a) top view, (b) side view, (c) front view

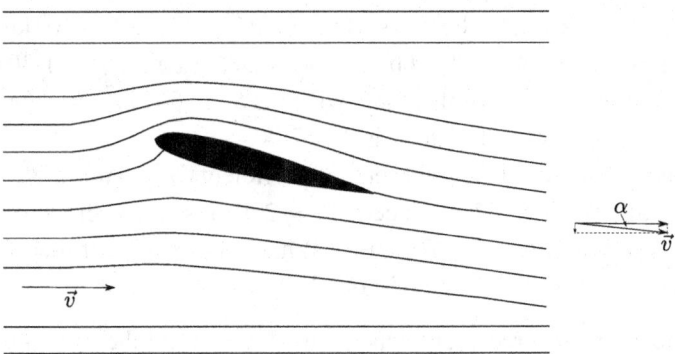

Figure 2.28 The scheme of air flow near the plane wing

By adjusting the position of the wing with respect to the direction of the flight the pilot can control the air flow around the wing and hence the interaction forces between the air and the wings. We consider the phenomenon in the reference frame of the airplane. The speed of air incident on the wings is v and it is directed horizontally. The shape and the position of the wings are such that the speed of air after the passage around the wings is still v but the angle between its direction and horizontal is a small angle α (Figure 2.28).

(a) Determine the vertical component F_L of the force by which the air acts on the plane (lift force). In your answer express F_L in terms of ρ, b, h, v, and α.
(b) Determine the horizontal component F_R of the force by which the air acts on the plane (drag force). What is the power of force F_R? Express your results in terms of the same quantities as in part (a).

The lift and the drag force are given by the expressions:

$$F_L = \frac{1}{2} C_L \rho S_w v^2 \qquad (2.98)$$

and

$$F_R = \frac{1}{2} C_R \rho S_w v^2, \qquad (2.99)$$

where C_L and C_R are dimensionless coefficients, while $S_w = 2bc$ is the cross-section of the airplane wings in the plane perpendicular to the flight direction.

(c) Express the coefficients C_L and C_R in terms of the angle α and the ratio of the wingspan and the wing thickness $A = 2b/c$.
(d) The pilot of a Boeing 747 set the position of the wings during takeoff so that $C_L = 1.50$. What is the angle α in that case? Determine the speed of the plane

that enables the takeoff. The mass of the plane is $M = 3.90 \cdot 10^5$ kg, the gravitational acceleration is $g = 9.81$ m/s^2, the density of air is $\rho = 1.30$ kg/m^3, the area of the cross-section of the plane wings is $S_w = 541$ m^2, and the ratio of the wingspan and the wing thickness is $A = 7.00$.

(e) To which values should the pilot set the coefficient C_L and the angle α to enable the cruise of a Boeing 747 at a speed of $v = 220$ m/s and a height of 5 km where the density of air is $\rho = 0.700$ kg/m^3? What is the power of drag force in that case? Neglect the dependence of g on elevation.

In the helicopter lift force occurs due to air flow around the rotating blades. We consider the helicopter shown in Figure 2.29 that has two rotating blades the frequency of rotation of which is f. The length of each blade is b, the thickness of the blades is c (Figure 2.29), while the lift coefficient is C_L. Assume that the helicopter is at rest and that the air flow around the blades occurs only due to their rotation.

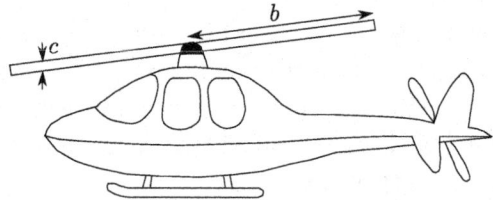

Figure 2.29 The scheme of the helicopter. b is the length of each blade while c is their thickness

(f) Express the lift force acting on the helicopter in terms of the quantities C_L, c, b, f, and the air density ρ.

(g) What is the frequency of the blades rotation that enables the helicopter to hover? The helicopter mass is $M = 4,500$ kg, $C_L = 0.500$, $\rho = 1.30$ kg/m^3, $c = 0.136$ m, $b = 7.30$ m.

Solution of Problem 12

(a) In time Δt the air that interacts with the wings changes the vertical component of the momentum by $\Delta p_y = \Delta m v_y$, where v_y is the vertical component of the air velocity after the passage around the wings, and Δm is the mass of the air that in time Δt passes through the cross-section of area

$$S = \pi b h. \tag{2.100}$$

Since

$$\Delta m = \rho S v \Delta t, \tag{2.101}$$

$$v_y = v\sin\alpha \approx v\alpha \tag{2.102}$$

and $F_L = \frac{\Delta p_y}{\Delta t}$ (Newton's second law), we obtain:

$$F_L = \rho\pi bhv^2\alpha. \tag{2.103}$$

(b) In time Δt the air that interacts with the wings changes the horizontal component of the momentum by $\Delta p_x = \Delta mv(1 - \cos\alpha)$. Since $F_R = \frac{\Delta p_x}{\Delta t}$ (Newton's second law), and $1 - \cos\alpha \approx \alpha^2/2$ for small angle α, using equations (2.100) and (2.101) we obtain

$$F_R = \frac{1}{2}\rho\pi bhv^2\alpha^2. \tag{2.104}$$

The power of F_R is

$$P_R = F_R v = \frac{1}{2}\rho\pi bhv^3\alpha^2. \tag{2.105}$$

(c) From equations (2.103), (2.98), and (2.97) we obtain

$$C_L = \frac{2\pi\alpha}{1+2/A}, \tag{2.106}$$

while from equations (2.104), (2.99), and (2.97) we have

$$C_R = \frac{\pi\alpha^2}{1+2/A}. \tag{2.107}$$

(d) From equation (2.106) it follows that

$$\alpha = (1+2/A)\frac{C_L}{2\pi} = 17.6°. \tag{2.108}$$

The plane can take off when the magnitude of the lift force becomes equal to the magnitude of the gravity force, hence

$$Mg = \frac{1}{2}C_L\rho S_w v^2, \tag{2.109}$$

and consequently $v = \sqrt{\frac{2Mg}{C_L \rho S_w}} = 85.2 \frac{m}{s}$.

(e) From equation (2.109) it follows that $C_L = \frac{2Mg}{\rho S_w v^2} = 0.417$. Using equation (2.108) we obtain $\alpha = 4.89°$. The power of the drag force is $P_R = F_R v$. Therefore, from equations (2.99) and (2.107) it follows that $P = \frac{1}{2}\frac{\pi\alpha^2}{1+2/A}S_w\rho v^3 = 36.0\,\text{MW}$.

(f) To determine the lift force acting on the helicopter we add the lift forces that act on each part of the blades. The area of the part of the length dr that is at a distance r from the axis of the propeller is $dS = cdr$, and the speed of that part is $v(r) = 2\pi f r$. The lift force acting on that part is $dF = \frac{1}{2}C_L\rho dS v(r)^2$. The total lift force acting on the helicopter propeller is $F = 2\int_0^b dF$, where the factor 2 originates from the fact that the helicopter has two blades. Finally $F = C_L\rho c 4\pi^2 f^2 \int_0^b r^2 dr = \frac{4\pi^2}{3}C_L c\rho f^2 b^3$.

(g) The helicopter hovers when the magnitude of the gravity force is equal to the magnitude of the lift force $Mg = \frac{4\pi^2}{3}C_L c\rho f^2 b^3$. One then obtains $f = \sqrt{\frac{3Mg}{4\pi^2 C_L b^3 c\rho}} = 9.88\,\text{Hz}$.

We refer the reader interested in physics of flight to reference [21].

Problem 13   Rocket

How much fuel is needed to accelerate a rocket to the first cosmic velocity? What is going on inside the rocket?

Most rockets use some liquid as propulsion fuel. The scheme of one such rocket is given in Figure 2.30. A chemical reaction occurs in a combustion chamber. This reaction transforms liquid fuel into a high-temperature mixture of gases. A jet of gases outflows through the nozzle. Consequently the rocket accelerates. We investigate in this problem the amount of fuel needed to accelerate the rocket to the first cosmic velocity so that it becomes the satellite of the Earth, and the efficiency of usage of chemical energy of the fuel in this case.

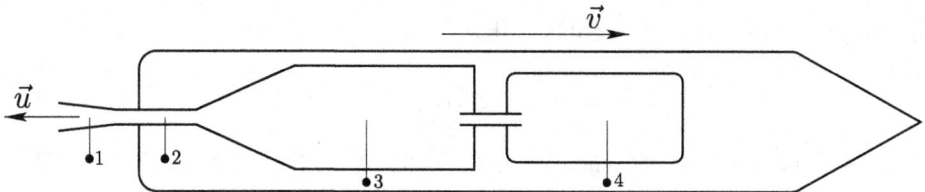

Figure 2.30 Scheme of the rocket: 1 – nozzle, 2 – throat, 3 – combustion chamber, 4 – reservoir. Velocity of the rocket is $\vec{v}$, while $\vec{u}$ is the velocity of the mixture of gases at the exit of the nozzle in the reference frame of the rocket

(a) Determine the speed of the satellite on a low-altitude orbit around the Earth. The radius of a low-altitude orbit is approximately equal to the radius of the Earth. What is the kinetic energy of this satellite if its mass is $m_s = 200\,\text{kg}$? Gravitational acceleration is $g = 9.81\,\frac{\text{m}}{\text{s}^2}$, while the radius of the Earth is $R_z = 6{,}400\,\text{km}$.

(b) The mixture of gases outflows at speed u in the reference frame of the rocket. The rocket was initially at rest. Then its mass was m_0. What will be the speed of the rocket v when its mass is m? Neglect the influence of Earth's gravity and atmosphere on the motion of the rocket.

(c) Liquid nitromethane (CH_3NO_2) is used as the fuel in the rocket. Nitromethane is kept in the reservoir at temperature $T_r = 300\,K$ and pressure p. It is injected into the combustion chamber from the reservoir. A chemical reaction at a constant pressure p takes place in the chamber,

$$CH_3NO_2 \longrightarrow 0.8\,CO + 0.2\,CO_2 + 0.7\,H_2 + 0.8\,H_2O + 0.5\,N_2,$$

which converts liquid nitromethane into a mixture of gases. The amount of heat released in the reaction of one mole of nitromethane at temperature T_r and pressure p is $q_m = 246\,\frac{kJ}{mol}$. The mixture of gases can be considered as an ideal gas with an adiabatic index of $\gamma = 1.25$ and molar mass $M = 20.3\,\frac{g}{mol}$. Calculate the temperature T of the mixture of gases. Assume that all heat released in the reaction is used to heat the mixture of gases at a constant pressure p. The gas constant is $R = 8.31\,\frac{J}{mol \cdot K}$.

(d) The difference of pressures in the chamber and outside the rocket causes the flow of the gas mixture through the throat and the nozzle and its final outflow from the rocket. Find the speed u of the outflow of the gas mixture from the rocket (in the reference frame of the rocket) if the temperature of the mixture in the chamber is T. Assume that the cross-section of the nozzle is much smaller than the cross-section of the chamber and that the exterior of the rocket is in a vacuum. Calculate the numerical value of u when the fuel from part (c) is used.

(e) What is the mass of nitromethane that should be used to accelerate the rocket from rest to the first cosmic velocity when the final mass of the rocket is $m_s = 200\,kg$?

Efficiency of usage of the chemical energy of the fuel is defined as the ratio of the final kinetic energy of the rocket and the heat released during the combustion of the liquid fuel that is used.

(f) If the term containing T_r is neglected in part (c) (i.e., $T_r = 0$ is put in that expression), the efficiency can be expressed only in terms of the ratio v/u of the speed of the rocket and the speed of fuel outflow. Determine the mathematical expression that describes this dependence. What is the efficiency for the rocket considered in this problem?

(g) For which v/u is the efficiency maximal and what is the maximal efficiency? You can determine the numerical values in this part of the problem using graphical or numerical techniques.

Solution of Problem 13

(a) Equation of motion of the satellite on the orbit yields $\frac{m_s v^2}{R_z} = m_s g$, where v is the speed of the satellite. This implies $v = \sqrt{gR_z} = 7.92\,\frac{\text{km}}{\text{s}}$, while the kinetic energy is $T = \frac{1}{2}m_s v^2$, i.e. $T = \frac{1}{2}m_s g R_z = 6.28 \cdot 10^9\,\text{J}$.

(b) Conservation of momentum at times t and $t + dt$ gives

$$m(t)v(t) = m(t+dt)v(t+dt) + [m(t) - m(t+dt)][v(t+dt) - u], \quad (2.110)$$

where the term on the left-hand side is the momentum of the rocket at time t, the first term on the right-hand side is the momentum of the rocket at time $t + dt$, and the second term on the right-hand side is the momentum of the gas ejected in the time interval $(t, t + dt)$. We obtain next

$$[m(t) - m(t+dt)]u = m(t)[v(t+dt) - v(t)], \quad (2.111)$$

which leads to $-\frac{dm}{m} = \frac{1}{u}dv$. Integration of the last equation yields $-\int_{m_0}^{m} \frac{dm'}{m'} = \frac{1}{u}\int_0^v dv'$, and consequently $v = u \cdot \ln\frac{m_0}{m}$. This equation is known as the Tsiolkovsky equation.

(c) When n_L moles of the reactant participate in the reaction, $n_D = 3n_L$ moles of gas mixture are created. Since the heat released heats the gas mixture, it follows that $q_m n_L = n_D C_p(T - T_r)$, where $C_p = \frac{\gamma}{\gamma-1}R$ is the molar heat capacity of the gas mixture at constant pressure. Hence we obtain $T = T_r + \frac{q_m}{R}\frac{\gamma-1}{\gamma}\frac{n_L}{n_D} = 2{,}270\,\text{K}$.

(d) Consider the part of gas that was at time t in the part of space that consists of the areas b and c, shown in Figure 2.31. Due to outflow of gas from the chamber, that part of gas is at time $t + \Delta t$ in the part of space that consists of areas a and b. We denote as Δm the mass of gas mixture that outflows in small time Δt, Δn is the number of moles corresponding to that mass, and ΔV is the volume of that amount of gas in the chamber, i.e. the volume of area c. We investigate the phenomenon in the reference frame of the rocket. The work performed by the rest of the system on the considered part of the gas is $\Delta A = p\Delta V$ since the pressure p acts from the right side on this part of the gas, while there is no pressure from the left side, which is in a vacuum. The change of kinetic energy of directed motion of this part of the gas is equal to the difference of the kinetic energy of the part of gas in area a at time $t + \Delta t$ and the part of gas in area c at time t, i.e., $\Delta T = \frac{1}{2}\Delta m \cdot u^2$. The change of internal energy of the considered part of the gas is also equal to the difference of internal energy of the part of gas in area a at time $t + \Delta t$ and the part of gas in area c at time t, which leads to $\Delta U = -\Delta n C_v T$. Conservation of energy gives $\Delta A = \Delta T + \Delta U$, i.e., $p\Delta V = \frac{1}{2}\Delta m u^2 - \Delta n C_v T$. Using the equation of state $p\Delta V = \Delta n RT$, as well as the equations $C_v = \frac{R}{\gamma-1}$ and $M = \frac{\Delta m}{\Delta n}$, we obtain $u = \sqrt{\frac{2\gamma}{\gamma-1}\frac{RT}{M}}$. We finally have $u = \sqrt{\frac{2q_m}{M}\frac{n_L}{n_D} + \frac{2\gamma}{\gamma-1}\frac{RT_r}{M}} = 3.05\,\frac{km}{s}$.

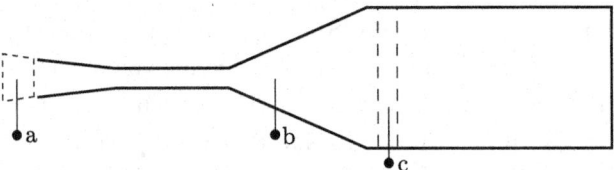

Figure 2.31 Schematic view of the outflow of gas from a combustion chamber

(e) From part (b) we have $m_s = m_0 \exp\left(-\frac{v}{u}\right)$. Therefore, the mass of the fuel used is $m_g = m_0 - m_s = m_s\left(\exp\frac{v}{u} - 1\right) = 2.49 \cdot 10^3\,\text{kg}$.

(f) The efficiency is $\eta = \frac{\frac{1}{2}m_s v^2}{m_g \frac{q_m}{M}\frac{n_L}{n_D}}$. Using the solutions of parts (d), (e), and (b) it follows that $\eta = \frac{(v/u)^2}{\exp\frac{v}{u}-1}$. Since (using $T_r = 0$) $v/u = 2.79$, it follows that $\eta = 0.51$.

(g) The graph of the function $f(x) = \frac{x^2}{e^x-1}$ is shown in Figure 2.32. We read from the graph that the maximum is reached for $x = 1.59$ and that this maximum is equal to $\eta = 0.648$.

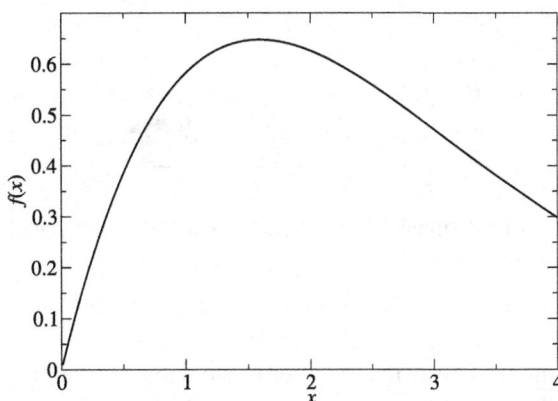

Figure 2.32 The graph of the function $f(x) = \frac{x^2}{e^x-1}$

We refer the reader interested in the physics and technology of rockets to reference [35]. This problem was given by the authors at the Serbian Physics Olympiad for high school students in 2012.

Problem 14   Global Positioning System

How much do the effects of relativity affect the operation of the Global Positioning System (GPS)?

The Global Positioning System (GPS) is a satellite navigation system that gives information about spatial and time coordinates and the velocity of an object on the Earth that is equipped with a GPS receiver. The system consists of 24 satellites that orbit at a height of $h = 20.00 \cdot 10^3$ km above the surface of the Earth. The period of rotation of the satellites is $T = 11\,\text{h}\,58\,\text{min}$. The radius of the Earth is $R_E = 6.371 \cdot 10^3$ km.

(a) Calculate the speed of the satellites v.

In this problem we see that the effects of the special theory of relativity (STR) are important for precise determination of the position of an object using GPS.

Consider a one-dimensional model of the GPS. A vehicle equipped with a GPS receiver is moving at speed u ($u \ll v$) in the direction of the x-axis of reference frame of the Earth S. Two satellites that emit a GPS signal are moving at speed v [calculated in part (a)] in the direction of the x-axis; see Figure 2.33. Satellite 1 is at the origin of reference frame S' of the satellites. The distance between the satellites in the system S' is $L = 2.400 \cdot 10^4$ km. At time $t = t' = 0$, the origins of the systems S and S' coincide.

Figure 2.33 One-dimensional model of the GPS system – the vehicle and the two satellites

(b) At certain moment of time, the receiver in the vehicle receives the signals emitted from satellites 1 and 2, respectively, at times t_1' and t_2' (the quantities marked with the prime sign refer to reference frame S', while those without it refer to reference frame S). Determine the expression for the x-coordinate of the vehicle (x_r) at that moment in terms of t_1', t_2', v, L, and the speed of light in vacuum c. Assume that $x_1 \leq x_r \leq x_2$ (where x_1 and x_2 are the coordinates of the satellites).
(c) Calculate the difference $\Delta x_{r-nr} = |x_r - x_{nr}|$ of the positions of the vehicle determined using STR (x_r) and using nonrelativistic mechanics (x_{nr}). The speed of light in a vacuum is $c = 2.998 \cdot 10^8 \, \frac{\text{m}}{\text{s}}$.

Note: If it is necessary in parts (c) and (f), when you consider the problem using nonrelativistic mechanics, assume that the speed of light in S is c while the speed of light in S' can be determined from the nonrelativistic law for velocity addition.

(d) Satellites are equipped with atomic clocks that enable determination of the moment of emission of the signal (in frame S') with precision of $\Delta t' = 1.0\,\mathrm{ns}$. Calculate the absolute error of determination of the x-coordinate of the vehicle that originates from the uncertainty in determination of time using the atomic clock.

The velocity of the vehicle is determined using the GPS system from the shift $\Delta f = f - f_0$ of the frequency f of the signal registered in the receiver with respect to the frequency f_0 of the signal emitted from the satellite, which is equal to $f_0 = 1.575\,\mathrm{GHz}$. The frequency shift of the signal emitted from satellite 1 and registered by the receiver is $\Delta f_1 = 20.30\,\mathrm{kHz}$.

(e) Determine the magnitude and the direction of the velocity of the vehicle.
(f) Calculate the difference $\Delta u_{\mathrm{r-nr}} = |u_{\mathrm{r}} - u_{\mathrm{nr}}|$ of the speeds of the vehicle determined from the frequency shift when the effects of STR are taken into account (u_{r}) and when these are not taken into account (u_{nr}).

Solution of Problem 14

(a) The magnitude of the angular velocity of the rotation of the satellite is $\omega = \frac{v}{h+R_E}$. Since $\omega = \frac{2\pi}{T}$, it follows that $v = \frac{2\pi}{T}(h + R_E)$ and consequently $v = 3.84619\,\frac{\mathrm{km}}{\mathrm{s}} \approx 3.846\,\frac{\mathrm{km}}{\mathrm{s}}$.

(b) Assume that the vehicle simultaneously received at time t the signals from both satellites, where the signal from the satellite i was emitted at time t_i, when the coordinate of that satellite was x_i ($i = 1, 2$). Since the speed of light is c, it follows that $x_{\mathrm{r}} - x_1 = c(t - t_1)$ and $x_2 - x_{\mathrm{r}} = c(t - t_2)$. From Lorentz transformations it follows that the coordinates x_i, t_i of the event "emission of the signal from satellite i" in the reference frame S are connected to corresponding coordinates x_i', t_i' in reference frame S' as $x_i = \dfrac{x_i' + vt_i'}{\sqrt{1 - \frac{v^2}{c^2}}}$, $t_i = \dfrac{t_i' + \frac{v}{c^2}x_i'}{\sqrt{1 - \frac{v^2}{c^2}}}$. Taking into account that $x_1' = 0$ and $x_2' = L$, it follows that the x-coordinate of the vehicle is $x_{\mathrm{r}} = \dfrac{\left(1 + \frac{v}{c}\right)L + \left(1 + \frac{v}{c}\right)ct_2' - \left(1 - \frac{v}{c}\right)ct_1'}{2\sqrt{1 - \frac{v^2}{c^2}}}$.

(c) The equations $x_{\mathrm{nr}} - x_1 = c(t - t_1)$ and $x_2 - x_{\mathrm{nr}} = c(t - t_2)$ also hold in nonrelativistic cases. From the Galilean transformations we have $x_i = x_i' + vt_i'$, $t_i = t_i'$. It then follows from previous equations that $x_{\mathrm{nr}} = \frac{L}{2} + \frac{1}{2}\left(1 + \frac{v}{c}\right)ct_2' -$

$\frac{1}{2}\left(1-\frac{v}{c}\right)ct_1'$. Since $v/c \ll 1$, one can use the approximation $\left(1-\frac{v^2}{c^2}\right)^{-1/2} \approx 1+\frac{1}{2}\frac{v^2}{c^2}$ in the expression for x_r, which leads to $x_r = x_{nr} + \frac{v}{2c}L+\ldots$, where the terms proportional to second or higher degree of small quantity v/c were omitted. Consequently we have $\Delta x_{r-nr} = \frac{v}{2c}L$, which leads to $\Delta x_{r-nr} = 154.0\,\mathrm{m}$. This result implies that the effects of STR are important for accurate determination of the position of the vehicle.

(d) Absolute error in determination of the coordinate that originates from uncertainty in the measurement of time using the atomic clock is $\Delta x_r = \frac{1}{2\sqrt{1-\frac{v^2}{c^2}}}[(c+v)\Delta t_2' + (c-v)\Delta t_1'] = \frac{c\Delta t'}{\sqrt{1-\frac{v^2}{c^2}}} \approx c\Delta t'$, where we used $\Delta t_1' = \Delta t_2' = \Delta t'$. We then obtain $\Delta x_r = 30\,\mathrm{cm}$.

(e) From the relativistic formula for the Doppler effect, the frequency of the signal that the receiver registers is $f = f_0\frac{\sqrt{1-\frac{v_r^2}{c^2}}}{1-\frac{v_r}{c}}$, where v_r is the velocity of satellite 1 in the reference frame of the vehicle. Using $\Delta f_1 = f - f_0$, it follows that $v_r = c\frac{\left(1+\frac{\Delta f_1}{f_0}\right)^2 - 1}{\left(1+\frac{\Delta f_1}{f_0}\right)^2 + 1}$, which leads to $v_r = 3{,}864.06\,\frac{\mathrm{m}}{\mathrm{s}}$. The relativistic law of velocity addition yields $v_r = \frac{v-u_r}{1-\frac{vu_r}{c^2}}$, where u_r is the x-component of the velocity of the vehicle in the reference frame of the Earth. It follows from previous equations that $u_r = \frac{v-v_r}{1-\frac{vv_r}{c^2}} = -17.874\,\frac{\mathrm{m}}{\mathrm{s}}$. Consequently the magnitude of the velocity of the vehicle is $17.874\,\frac{\mathrm{m}}{\mathrm{s}}$, while its direction is opposite to the direction of the x-axis.

(f) From the nonrelativistic formula for the Doppler effect, the frequency of the signal that the receiver registers is $f = \frac{f_0}{1-\frac{v_r}{c_r}}$, where v_r is the velocity of satellite 1 in the reference frame of the vehicle, and c_r is the speed of light in that reference frame. The non-relativistic law of velocity addition yields $v_r = v - u_{nr}$ and $c_r = c - u_{nr}$, where u_{nr} is the x-component of the velocity of the vehicle in reference frame of the Earth. Using $\Delta f_1 = f - f_0$, we obtain from previous equations $u_{nr} = v\left(1+\frac{\Delta f_1}{f_0}\right) - c\frac{\Delta f_1}{f_0} = -17.849\,\frac{\mathrm{m}}{\mathrm{s}}$. The difference in question is $\Delta u_{r-nr} = |u_r - u_{nr}| = 2.5\,\frac{\mathrm{cm}}{\mathrm{s}}$. We therefore conclude that the speed of the vehicle can be quite accurately determined even without taking relativistic effects into account.

This problem is a combination of two problems given by Veljko Janković at the national physics competition for the fourth grade of high school in Serbia in 2018.

3

Sport

Problem 15   How Far Can Athletes Jump?

What are the world records in the pole vault, the high jump, and the long jump?

We investigate in this problem how long or how high athletes can jump using different techniques. The mass of an athlete is $m = 75.0\,$kg. The position of the athlete's center of mass is $h_{CM} = 0.900$ m above the ground. Gravitational acceleration is $g = 9.81\,$m/s^2. Unless otherwise stated, neglect air resistance.

(a) We consider the pole vault first. In the course of a pole vault the athlete first runs and reaches a speed of $v_1 = 10.0\,$m/s. The athlete then uses an elastic pole to efficiently convert kinetic energy into potential energy (Figure 3.1). In pole vault competitions, the goal is to jump above a bar that is in a horizontal position at a height of h_b above the ground. Assume that the speed of the athlete in the moment of passing above the bar is negligible (which is a justified assumption since the athlete can push off the pole in that moment and therefore does not need a significant speed to pass above the bar) and that the center of mass in that moment is $\Delta h = 0.100$ m above the bar. What will be the height h_b of this jump if the total kinetic energy that the athlete had at the end of his/her run-up was converted into potential energy in the moment when the athlete passes above the bar?

In the long jump, the athlete runs, reaching speed v_0, and then jumps as shown in Figure 3.2.

(b) Find the maximal length of the jump for an athlete who has reached a speed of $v_0 = 10.0\,$m/s during the run-up. Assume that the athlete does not lose kinetic energy at the moment of takeoff and that the center of mass is at the same height

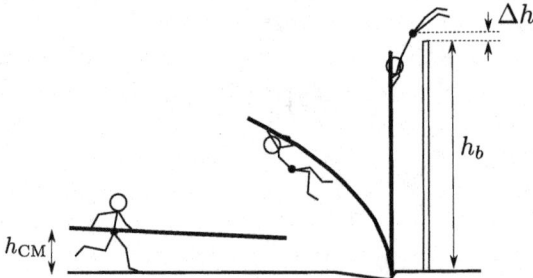

Figure 3.1 Three characteristic positions of an athlete during the pole vault. The dark circle denotes the position of the center of mass

in the moment of landing as in the moment of takeoff. What is the highest position of the athlete's center of mass during the jump in which the maximal length is reached?

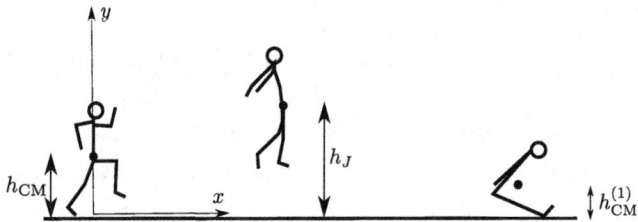

Figure 3.2 Positions of an athlete during a long jump. The dark circle denotes the position of the center of mass

The length and the height obtained in part (b) are larger than the world records in the long jump and the high jump. Therefore, the assumptions introduced are not fully realistic.

The assumption that the kinetic energy of the athlete does not change while the velocity changes its direction at takeoff is not realistic. For this to happen, a significant force must act on the athlete vertically. It is not easy to produce such a force.

(c) To establish the amount of kinetic energy that the athlete can produce at takeoff, we consider a vertical high jump without the run-up. During this jump the athlete starts from a squatting position pushing the feet off the ground. At the moment when the feet leave the ground, the athlete is in a vertical position, as shown in Figure 3.3. If the athlete jumped to a height of $h_2 - h_1 = 90.0\,\text{cm}$ (Figure 3.3), what was the athlete's kinetic energy immediately after takeoff?

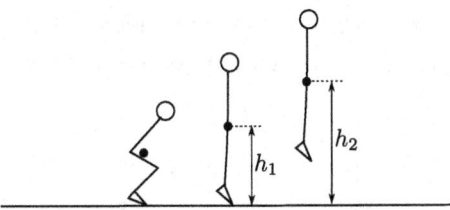

Figure 3.3 Three characteristic positions of an athlete during a vertical high jump. The dark circle denotes the position of the center of mass

(d) We consider next an athlete competing in the long jump who reaches a speed of $v_0 = 10.0\,\text{m/s}$ during the run-up. Assume that the athlete increases the kinetic energy during takeoff by the amount determined in the previous part of the problem. The athlete accomplishes this by producing the vertical component of the velocity, while the horizontal component remains the same. To jump longer, the jumper changes the position of the body before landing so that the center of mass is $h_{\text{CM}}^{(1)} = 0.700\,\text{m}$ above the ground in the moment of landing (Figure 3.2). What is the length of the jump under these assumptions?

(e) The magnitude of the air drag force is

$$F_D = \frac{1}{2}C_D\rho A v^2, \tag{3.1}$$

where $\rho = 1.225\,\text{kg/m}^3$ is the density of air, $A = 0.500\,\text{m}^2$ is the surface of the cross-section of the athlete, $C_D = 0.900$ is the dimensionless drag coefficient, and v the speed. For simplicity assume that the direction of the drag force is horizontal and that its magnitude is constant and equal to the magnitude before takeoff. How smaller will be the result obtained in part (d) when the drag force is taken into account?

(f) In this part of the problem we consider the high jump (Figure 3.4). The high jump athletes take their run-up on a curved path. For this reason, at the moment of takeoff they are inclined with respect to the horizontal (see Figures 3.4 and 3.5). At the moment just before the jumper stood on the take-off foot, the jumper's velocity was horizontal and its magnitude was $v_0 = 8.5\,\text{m/s}$ (Figure 3.5). You can assume that in this moment the athlete can be approximated as a homogeneous, rigid rod of length $2l = 2.00\,\text{m}$ directed at an angle θ with respect to the horizontal. Find the dependence of the vertical component of the athlete's velocity immediately after the athlete stands on the take-off leg on angle θ and speed v_0. Next, while his/her foot is still in contact with the ground, the athlete increases the vertical component of the velocity and therefore increases the kinetic energy by the amount determined in part (c). What will be the height of the jump if the highest point of the trajectory of the athlete's center of mass is $\Delta h = 5.00\,\text{cm}$ above the bar? Plot the graph of the dependence of the height of the jump on θ and find the maximal height of the jump using the graph.

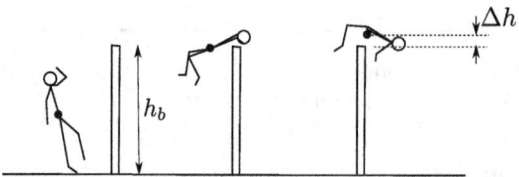

Figure 3.4 Three characteristic positions of an athlete during a high jump. The dark circle denotes the position of the center of mass

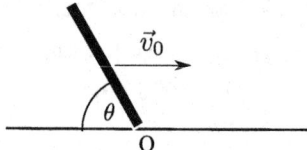

Figure 3.5 The model of an athlete competing in the high jump

Solution of Problem 15

(a) Conservation of energy yields $\frac{1}{2}mv_1^2 + mgh_{CM} = mg\,(h_b + \Delta h)$, implying $h_b = h_{CM} - \Delta h + \frac{v_1^2}{2g} = 5.90\,\text{m}$. This height is somewhat shorter than the pole vault world record for men of 6.20 m set by Armand Duplantis from Sweden on March 20, 2022, while it is larger than the women's world record of 5.06 m set by Yelena Isinbayeva from Russia on August 28, 2009.

(b) We define α as the angle between the athlete's velocity and the horizontal at the moment immediately after takeoff. The dependence of the coordinates of the center mass on time is:

$$x(t) = v_0 t \cos\alpha, \tag{3.2}$$

$$y(t) = h_{CM} + v_0 t \sin\alpha - \frac{1}{2}gt^2. \tag{3.3}$$

At the moment of landing we have $y(t_d) = h_{CM}$. Using equation (3.3) it follows that $t_d = \frac{2v_0 \sin\alpha}{g}$. By putting this expression in equation (3.2) and using the trigonometric identity $\sin(2\alpha) = 2\sin\alpha\cos\alpha$ we find that the length of the jump is $D = x(t_d) = v_0^2 \frac{\sin(2\alpha)}{g}$. This expression is maximal when $\sin(2\alpha) = 1$;

that is, $\alpha = 45°$. The maximum is equal to $D_m = \frac{v_0^2}{g} = 10.2\,\text{m}$. The athlete reaches the highest point of the trajectory in the moment $t_1 = \frac{v_0 \sin \alpha}{g}$. In that moment, using equation (3.3), we find $y_m = y(t_1) = h_{CM} + \frac{v_0^2 \sin^2 \alpha}{2g}$ – that is, $y_m = h_{CM} + \frac{v_0^2}{4g} = 3.45\,\text{m}$. The jump is longer than the women's long jump world record of $7.52\,\text{m}$ set by Galina Chistyakova from the Soviet Union on June 11, 1988. It is also longer than the men's long jump world record of $8.95\,\text{m}$ set by Mike Powell from the USA on August 30, 1991. It is noticeable as well that the highest position of the center of mass is higher than the world record in the high jump.

(c) Let T be the kinetic energy of the athlete immediately after takeoff. Conservation of energy leads to $T + mgh_1 = mgh_2$, where h_1 is the position of the center of mass at the moment of takeoff, while h_2 is the highest position of the center of mass. We then find $T = mg\,(h_2 - h_1) = 662\,\text{J}$.

(d) The time from takeoff to the moment of reaching the highest point in the trajectory is $t_a = \sqrt{\frac{2(h_J - h_{CM})}{g}}$, while the time from reaching this point to landing is $t_b = \sqrt{\frac{2\left(h_J - h_{CM}^{(1)}\right)}{g}}$. The velocity of the jumper has only the horizontal component v_0 at the highest point of the trajectory. Conservation of energy applied to the moment immediately after takeoff and the moment when the athlete is in the highest point of the trajectory yields $\frac{1}{2}mv_0^2 + T + mgh_{CM} = \frac{1}{2}mv_0^2 + mgh_J$ – that is, $h_J = h_{CM} + \frac{T}{mg}$. The length of the jump is $D = v_0\,(t_a + t_b)$ – that is,

$$D = v_0 \cdot \left(\sqrt{\frac{2\,(h_J - h_{CM})}{g}} + \sqrt{\frac{2\left(h_J - h_{CM}^{(1)}\right)}{g}} \right) = 9.02\,\text{m}. \qquad (3.4)$$

The result is very close to the men's long jump world record, while it is longer than the world record for females.

(e) The drag force decelerates the horizontal component of the velocity by $a = \frac{F_D}{m} = \frac{C_D \rho A v_0^2}{2m}$. The duration of the jump is the same as in part (e) and is equal to $t = t_a + t_b = 0.902\,\text{s}$. The length of the jump when the drag force is taken into account is $D_1 = v_0 t - \frac{1}{2}at^2$. Therefore, the length of the jump is in this case smaller by $\Delta d = \frac{1}{2}at^2 = \frac{C_D \rho A v_0^2 t^2}{4m} = 0.15\,\text{m}$.

(f) The angular momentum of the athlete with respect to point O is $L_i = mv_0 l \sin \theta$ before the take-off leg touches the ground. During takeoff, the foot is in contact with point O. Therefore, it can be assumed that the athlete, modeled as a rigid rod, rotates around point O. The angular momentum of the jumper in this moment of time was $L_f = I\omega$, where $I = \frac{1}{3}m\,(2l)^2$ is the moment of inertia of the rod with respect to the axis that passes through its end, the direction of which coincides with the direction of the angular velocity with magnitude ω. Angular momentum conservation can be applied to the moments just before and just after the contact of the foot with point O, which reads $L_i = L_f$. The reason for this is that it is only the force mg that has the torque with respect to O, while the effect of this torque during a short period of time is negligible. We therefore find $l\omega = \frac{3}{4}v_0 \sin\theta$. The vertical component of the velocity of the rod is $v_v = l\omega \cos\theta$, which leads to $v_v = \frac{3v_0}{4}\sin\theta\cos\theta$. The kinetic energy of translation of the athlete immediately after takeoff is $T' = T + \frac{1}{2}mv_v^2$. Application of the energy conservation law to the moment immediately after takeoff and the moment when the center of mass is at the highest point gives $T' + mgl\sin\theta = mg(h_b + \Delta h)$. The previous two equations yield $h_b = l\sin\theta + \frac{T}{mg} - \Delta h + \frac{9v_0^2}{32g}\sin^2\theta\cos^2\theta$. This dependence is shown in Figure 3.6. We read from the graph that the maximal height of the jump is $h_b = 2.13$ m and it is reached for $\theta = 53.7°$. This value is smaller than the men's high jump world record of 2.45 m set by Javier Sotomayor from Cuba on July 27, 1993, while it is slightly larger than the women's high jump world record of 2.09 m set by Stefka Kostadinova from Bulgaria on August 30, 1987.

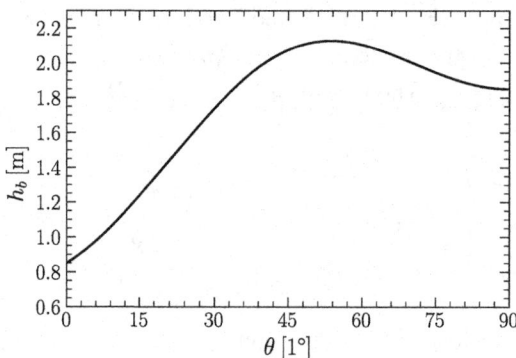

Figure 3.6 The dependence of the height of the jump on angle θ

We refer interested readers to references [10], [15], and [16].

Problem 16 Bungee Jumping

How should one connect two elastic cords if couples want to jump together? Is the bungee jumper in danger when exposed to large vertical accelerations?

Bungee jumping is one of the extreme disciplines practiced by adventure tourists. It involves a person jumping from a high object such as a building, a bridge, or a crane. The jumper is connected to the fixed object at the initial position via a long, elastic cord. The jumper is initially in the vertical position and jumps downward with no initial speed. At the lowest point of the jump, the jumper is in the horizontal position, as shown in Figure 3.7.

In this problem we consider a bungee jump from a crane at a height $h = 55.0\,\mathrm{m}$ above the surface of the water at Ada Ciganlija in Belgrade, as shown in Figure 3.7. The length of the elastic cord when it is not deformed is $l = 28.0\,\mathrm{m}$, while its stiffness is $k = 180\,\mathrm{N/m}$. Assume that the mass of the jumper is concentrated in the center of mass – that is, at height $h_{\mathrm{cm}} = 1.00\,\mathrm{m}$ above the surface where the jumper is standing. The mass of the elastic cord is negligible. Gravitational acceleration is $g = 9.81\,\mathrm{m/s^2}$. Neglect air resistance.

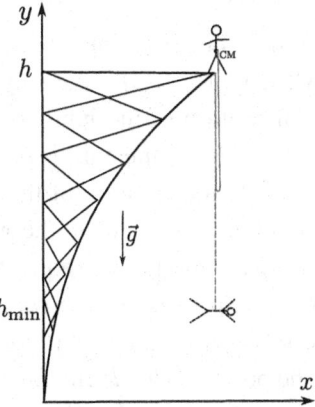

Figure 3.7 Crane for bungee jumping at Ada Ciganlija

(a) What is the maximal speed gained during the motion of a jumper of mass $m = 80.0\,\mathrm{kg}$?
(b) What is the minimal height above the surface of water reached by this jumper during the motion?
(c) What is the maximal possible mass of the jumper that will not fall into the water during the motion?
(d) It recently has become popular for couples to jump together. In this case, the elastic cord is doubled – that is, another cord of the same length l (when it is not deformed) is connected in parallel to the first cord. What is the minimal stiffness of the second cord that enables a safe jump of a boyfriend of mass $m_1 = 90.0\,\mathrm{kg}$ and his girlfriend of mass $m_2 = 60.0\,\mathrm{kg}$?
(e) Plot the graph of the dependence of the position of the jumper of mass $m = 80.0\,\mathrm{kg}$ on time. What is the period of oscillations of the jumper?

(f) Plot the graph of the time dependence of the y-component of the velocity and the speed of the jumper from part (e).

(g) Plot the graph of the time dependence of the y-component of acceleration and the magnitude of acceleration of the jumper from part (e).

(h) Large vertical accelerations that last longer than $\Delta t = 2\,\mathrm{s}$ lead to a strong effect of inertial forces on the flow of blood in the human body. This can cause serious consequences to the jumper. The largest safe magnitude of acceleration when it comes to the flow of blood to the legs is $5g$, while this value is $2g$ when the flow of blood to the head is concerned. Is the bungee jumper from part (e) in danger?

Solution of Problem 16

(a) We denote the position of the jumper at time t as $y(t)$. While the elastic cord is not tightened – that is, while $y(t) \geq h - l$ – the only force acting on the jumper is gravity. The equation of motion of the jumper in this case is $m\vec{a} = m\vec{g}$. In the opposite case, when $y(t) \leq h - l$, the cord is tightened and the elastic force acts on the jumper. The equation of motion of the jumper is then $m\vec{a} = m\vec{g} - k(y(t) - h + l)\vec{j}$, where $\vec{j}$ is the unit vector in the y-direction. In this case, the jumper performs harmonic oscillations. Consequently the motion of the jumper is periodic and consists of three phases: the jumper falls freely from $y = h + h_{cm}$ to $y = h - l$; then he performs a harmonic motion, first downward and then upward until he reaches the point $y = h - l$; and finally moves vertically upward under the action of gravity before reaching the maximal height. These phases then repeat periodically.

The speed of the jumper is maximal in equilibrium position y_0, which can be determined from the condition of equality of gravity and the elastic force $mg = k(h - l - y_0)$, which leads to $y_0 = h - l - mg/k$. Conservation of energy gives $mg(h + h_{cm}) = mgy_0 + \frac{1}{2}mv_{max}^2 + \frac{1}{2}k(h - l - y_0)^2$, which, along with the expression for y_0 yields $v_{max} = \sqrt{2g(l + h_{cm}) + mg^2/k} = 24.7\,\mathrm{m/s}$.

(b) Conservation of energy yields

$$mg(h + h_{cm}) = mgh_{min} + \frac{1}{2}k(h - l - h_{min})^2. \qquad (3.5)$$

The physically meaningful solution of this quadratic equation is $h_{min} = h - l - \frac{mg}{k} - \sqrt{\frac{m^2g^2}{k^2} + \frac{2mg(l+h_{cm})}{k}} = 6.15\,\mathrm{m}$.

(c) The maximal mass of the jumper that can jump without falling into the water can be obtained from equation (3.5) with the condition $h_{min} = 0$, which leads to $m_{max} = \frac{k(h-l)^2}{2g(h+h_{cm})} = 119\,\mathrm{kg}$.

(d) The equivalent stiffness of the cord is $k_e = k + k'$. From equation (3.5) and the condition $h_{min} = 0$ we obtain $k_e = \frac{2(h+h_{cm})(m_1+m_2)g}{(h-l)^2}$ – that is, $k' = \frac{2(h+h_{cm})(m_1+m_2)g}{(h-l)^2} - k = 46.1\,\text{N/m}$.

(e) The jumper falls freely from height $h + h_{cm}$ until time $t_1 = \sqrt{2(l+h_{cm})/g}$ $= 2.43\,\text{s}$ when the cord becomes tightened. The position of the jumper during the free fall is $y(t) = h + h_{cm} - gt^2/2$. The motion of the jumper from time t_1 is harmonic. Therefore, the position of the jumper is $y(t) = y_0 - A\sin(\omega t + \varphi)$, where $A = y_0 - h_{min}$ is the amplitude of oscillations, $\omega = \sqrt{k/m}$ is the angular frequency of oscillations, and φ is the phase. The phase φ can be determined from $y(t_1) = h - l = y_0 - A\sin(\omega t_1 + \varphi)$, which leads to $\varphi = \arcsin(\frac{y_0-h+l}{A}) - \omega t_1 = -3.91\,\text{rad}$.

The jumper performs harmonic motion downward until time t_2, which is determined from $\sin(\omega t_2 + \varphi) = 1$, which leads to $t_2 = (\pi/2 - \varphi)/\omega = 3.66\,\text{s}$. Time t_2 is equal to half of a period of oscillation of the jumper. Consequently the period is $T = 2t_2 = 7.31\,\text{s}$. After time t_2 the jumper performs a harmonic motion upward until time $t_3 = 2t_2 - t_1 = 4.88\,\text{s}$. The jumper moves upward under the action of gravity from time t_3. Consequently $y(t) = h - l + v_0(t - t_3) - g(t - t_3)^2/2$, where $v_0 = \sqrt{2(h_{cm}+l)g}$. The jumper reaches the initial position at time $t_4 = t_3 + t_1 = T$. The graph in question is shown in Figure 3.8.

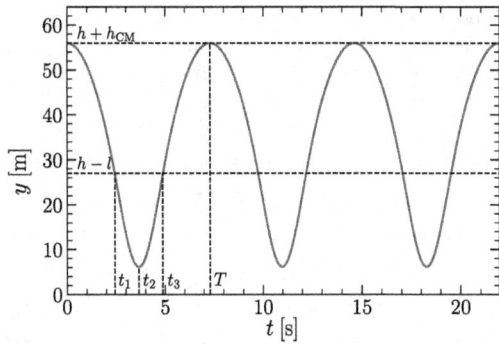

Figure 3.8 Time dependence of the position of the jumper

(f) The y-component of the velocity during the first period is

$$v(t) = \begin{cases} -gt & t \le t_1 \\ -A\omega\cos(\omega t + \varphi) & t_1 \le t \le t_3 \\ v_0 - g(t - t_3) & t_3 \le t \le t_4. \end{cases}$$

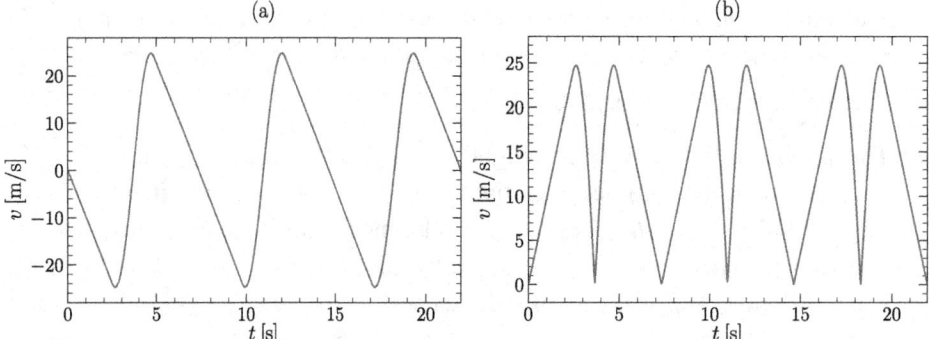

Figure 3.9 The graph of the time dependence of: (a) the *y*-component of the velocity and (b) the speed

Since the jumper is moving along the *y*-direction, the speed is equal to the absolute value of the *y*-component of the velocity. The graph in question is shown in Figure 3.9.

(g) The *y*-component of acceleration during the first period is

$$a(t) = \begin{cases} -g & t \leq t_1 \\ A\omega^2 \sin(\omega t + \varphi) & t_1 \leq t \leq t_3 \\ -g & t_3 \leq t \leq t_4. \end{cases}$$

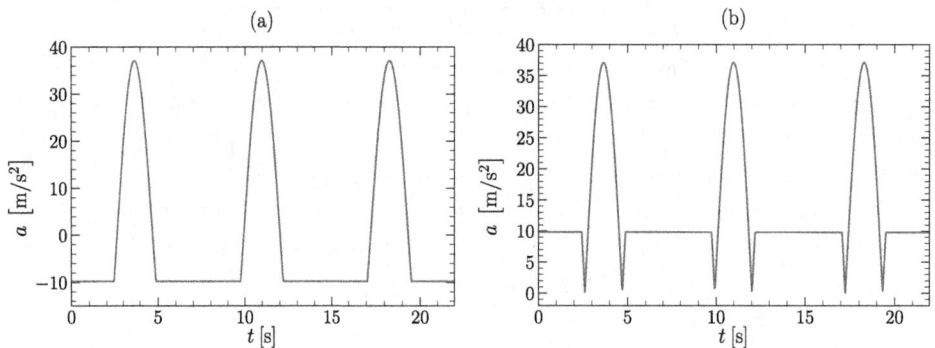

Figure 3.10 The graph of time dependence of: (a) the *y*-component of acceleration and (b) the magnitude of acceleration

The magnitude of the acceleration of the jumper is equal to the absolute value of the *y*-component of acceleration. The graph in question is shown in Figure 3.10.

(h) We see from the graph shown in Figure 3.11 that the magnitude of acceleration larger than $2g$ is present in the time interval that is shorter than $\Delta t = 2\,\mathrm{s}$. Consequently the jumper is not in danger.

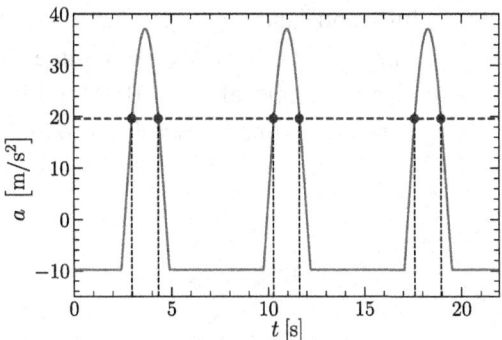

Figure 3.11 The graph of time dependence of the y-component of acceleration (solid line) and a safe magnitude of acceleration (horizontal dashed line)

Problem 17 Ski Jumping

How is the length of the jump affected by the wind and the in-run length? Does headwind decrease or increase the length of the jump?

Ski jumping is one of the most popular winter sports. A competitor in ski jumping first goes down the in-run hill (from point A to point B in Figure 3.12), takes off at the end of the in-run (point B in Figure 3.12), and then flies through the air. The aim is to land (the landing point is denoted as C in Figure 3.12) as far as possible from the point of takeoff.

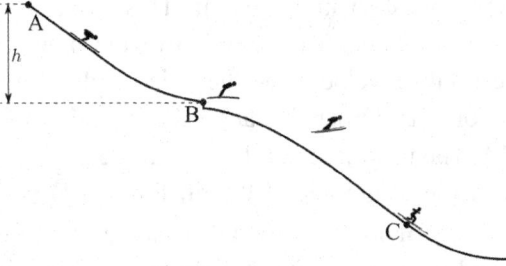

Figure 3.12 Scheme of a ski jumping hill and the positions of a ski jumper in different moments of time

For simplicity, you can assume in all parts of the problem that the ski jumper is a point particle of mass $m = 75.0\,\text{kg}$. The gravitational acceleration is $g = 9.81\,\frac{\text{m}}{\text{s}^2}$.

(a) The speed of the ski jumper at point B is $v_0 = 90.0\,\frac{\text{km}}{\text{h}}$. Determine the difference in altitude h of points A and B. In this part of the problem you can neglect the interaction of the skier with the air as well as the friction between the skis and the ground.

In reality, ski jumping hills and landing slopes have a complicated shape as shown in Figure 3.12. To simplify the analysis we assume that the in-run is an inclined plane with an inclination angle of $\gamma_0 = 10.0°$, while the landing slope is also an inclined plane with an inclination angle of $\delta = 38.0°$, as shown in Figure 3.13.

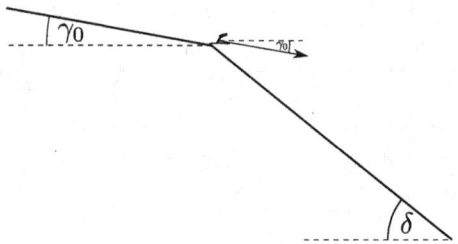

Figure 3.13 Simplified model of a ski jumping hill

(b) The angle between the direction of the skier's velocity and the horizontal is $\gamma_0 = 10.0°$ immediately after takeoff, as presented in Figure 3.13. Calculate the length of the jump by neglecting the interaction of the skier with the air. The length of the jump is defined as the distance between the place of takeoff and the place of landing.

The result obtained in the previous part of the problem does not give a good estimate of the length of the jump because the force of interaction between the skier and the air was not taken into account. This force has two components, the drag force and the lift force. The direction of the drag force is opposite to the direction of the relative velocity of the skier with respect to the air. The direction of the lift force is perpendicular to the direction of the drag force as shown in Figure 3.14. The magnitude of the drag force is given by the expression $F_D = \frac{1}{2}C_D S_{\text{eff}}\rho u^2$, and the magnitude of the lift force is $F_L = \frac{1}{2}C_L S_{\text{eff}}\rho u^2$, where $\rho = 1.17 \, \frac{\text{kg}}{\text{m}^3}$ is the density of air, $S_{\text{eff}} = 0.700 \, \text{m}^2$ is the area of the effective cross-section of the skier, and u is the magnitude of the relative velocity of the skier with respect to the air. The position of the skier's body is such that $C_D = 0.412$ and $C_L = 0.420$.

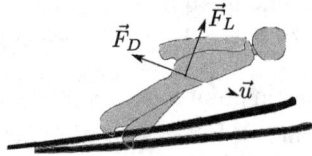

Figure 3.14 The directions of the drag force and the lift force

(c) Calculate the speed of the skier if a steady state in which the speed is constant was reached. Calculate also the angle between the skier's velocity and the
horizontal in this case.

The speed obtained in the previous part of the problem is much faster
than the initial speed of the skier, while the angle obtained is larger than the angle
of the landing slope. Therefore, the skier is not in a stationary state throughout most
of the jump. Consequently it is necessary to analyze how relevant kinematic quantities depend on time. In addition to velocity and acceleration, the relevant quantity
is the jerk, which is defined as the change of acceleration in a unit of time $\vec{j} = \frac{d\vec{a}}{dt}$.

(d) A particle is moving in such a way that the jerk $\vec{j}$ does not depend on time. The
position of the particle at time $t = 0$ is $\vec{r}_0$, its velocity is $\vec{v}_0$, and its acceleration
is $\vec{a}_0$. Determine the dependence of the position of the particle on time.

The motion of the skier is also affected by the presence of the wind, the velocity
of which is $\vec{w}$.

(e) Derive the expression for the jerk $\vec{j}$ in the moment immediately after takeoff.
The expression should contain the quantities C_D, C_L, m, ρ, S_{eff}, $\vec{v}_0$, $\vec{w}$, and $\vec{a}_0$
(and the relevant unit vectors).

To a good approximation, it can be assumed that the jerk of the skier is constant
during the jump and that it is equal to the jerk in the moment immediately after
takeoff.

(f) Calculate the length of the jump in the case when there is no wind.

Weather conditions can change during ski jumping competitions. For this reason, competitors who jump when the wind is favorable have the advantage. It can
also happen that in very favorable conditions the competitors jump distances longer
than the hill length, which can bring them into danger. In such cases, to avoid this
danger, the in-run length is shortened and the competition restarted. To avoid such
problems, the International Skiing Federation introduced in 2009 correction factors
that take into account the velocity of the wind and the in-run length. The competition can then be continued even when weather conditions change significantly. It
is therefore of significant interest to understand how wind and in-run length affect
the length of the jump.

(g) Calculate the length of the jump of the skier in the presence of a headwind with
speed $w = 3.00 \frac{\text{m}}{\text{s}}$, as well as in the case of a tailwind of the same magnitude.
The direction of the wind coincides with the direction of the landing slope.
Assume that the conditions of part (a) are fulfilled during the in-run.

(h) Calculate the length of the jump when there is no wind, but the difference in altitude of the beginning and the end of the in-run increases by $\Delta h = 0.500\,\text{m}$. Assume that the conditions of part (a) are fulfilled during the in-run.

Solution of Problem 17

(a) Conservation of energy in the moment when the skier is at the beginning of the in-run and the moment when the skier is taking off gives $mgh = \frac{1}{2}mv_0^2$, where h is the difference of the altitudes in question. This implies

$$h = \frac{v_0^2}{2g} = 31.855\,\text{m} \approx 31.9\,\text{m}. \tag{3.6}$$

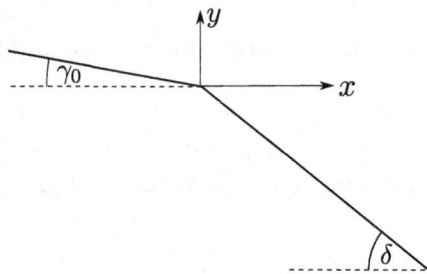

Figure 3.15 Simplified model of a ski jumping hill and the coordinate system used in the solution

(b) After takeoff the skier is moving at a constant acceleration equal to gravitational acceleration. We choose the coordinate system as in Figure 3.15. The x-component of the velocity of the skier is $v_x(t) = v_0 \cos \gamma_0$, while the y-component is $v_y(t) = -v_0 \sin \gamma_0 - gt$, where t is the time interval passed from takeoff. The dependence of the skier's coordinates on time is given by the expressions:

$$x(t) = v_0 t \cos \gamma_0, \tag{3.7}$$

$$y(t) = -v_0 t \sin \gamma_0 - \frac{1}{2}gt^2. \tag{3.8}$$

At the moment of landing, the coordinates satisfy the condition

$$\tan\delta = -\frac{y(t)}{x(t)}. \tag{3.9}$$

The length of the jump is

$$d = \frac{x(t)}{\cos \delta}, \tag{3.10}$$

where $x(t)$ is the coordinate of the skier in the moment of landing. From equations (3.7)–(3.10) we find

$$d = \frac{2v_0^2 \cos \gamma_0}{g \cos \delta} (\cos \gamma_0 \tan \delta - \sin \gamma_0) = 94.872\,\text{m} \approx 94.9\,\text{m}. \tag{3.11}$$

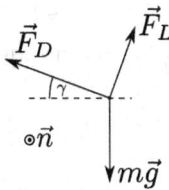

Figure 3.16 The forces acting on the skier during the jump

(c) The sum of all forces acting on the skier is zero in steady state. The forces acting on the skier during the jump are shown in Figure 3.16. Consequently we obtain $F_D = mg \sin \gamma$ and $F_L = mg \cos \gamma$, where γ is the angle between the skier's velocity and the horizontal, $F_D = \frac{1}{2}C_D S_{\text{eff}}\rho v^2$ is the magnitude of the drag force, and $F_L = \frac{1}{2}C_L S_{\text{eff}}\rho v^2$ is the magnitude of the lift force. We find from previous equations that $\tan\gamma = \frac{C_D}{C_L}$, which leads to $\gamma \approx 44.4°$, as well as

$$v = \sqrt{\frac{2mg}{\rho S_{\text{eff}}\sqrt{C_D^2+C_L^2}}} \quad \text{— that is, } v \approx 55.3\,\tfrac{\text{m}}{\text{s}}.$$

(d) The definition of the jerk implies $d\vec{a} = \vec{j}dt$. By performing the integration from the moment 0 to the moment t it follows that $\vec{a}(t) = \vec{a}_0 + \vec{j}t$. Next we have $\vec{v}(t) = \vec{v}_0 + \int_0^t du \cdot \vec{a}(u)$, which leads to $\vec{v}(t) = \vec{v}_0 + \vec{a}_0 t + \frac{1}{2}\vec{j}t^2$. Using $\vec{r}(t) = \vec{r}_0 + \int_0^t du \cdot \vec{v}(u)$, we obtain

$$\vec{r}(t) = \vec{r}_0 + \vec{v}_0 t + \frac{1}{2}\vec{a}_0 t^2 + \frac{1}{6}\vec{j}t^3. \tag{3.12}$$

(e) Newton's second law for the skier's motion reads

$$m\vec{a} = m\vec{g} + \vec{F}_D + \vec{F}_L. \tag{3.13}$$

By performing the differentiation we obtain

$$\vec{j} = \frac{1}{m}\left(\frac{d\vec{F}_D}{dt} + \frac{d\vec{F}_L}{dt}\right). \tag{3.14}$$

The drag force is given by the expression

$$\vec{F}_D = -\frac{1}{2}C_D\rho S_{\text{eff}}u\vec{u}, \tag{3.15}$$

where $\vec{u} = \vec{v} - \vec{w}$. After the differentiation we get

$$\frac{\mathrm{d}\vec{F}_D}{\mathrm{d}t} = -\frac{1}{2}C_D\rho S_{\text{eff}}\left(\frac{\mathrm{d}u}{\mathrm{d}t}\vec{u} + u\frac{\mathrm{d}\vec{u}}{\mathrm{d}t}\right). \tag{3.16}$$

Next we have

$$\frac{\mathrm{d}u}{\mathrm{d}t} = \frac{\mathrm{d}}{\mathrm{d}t}\left(\sqrt{\vec{u}\cdot\vec{u}}\right) = \frac{1}{2u}\frac{\mathrm{d}(\vec{u}\cdot\vec{u})}{\mathrm{d}t} = \frac{\vec{u}}{u}\frac{\mathrm{d}\vec{u}}{\mathrm{d}t}. \tag{3.17}$$

Taking into account that $\vec{a} = \frac{\mathrm{d}\vec{u}}{\mathrm{d}t}$, previous equations lead to

$$\frac{\mathrm{d}\vec{F}_D}{\mathrm{d}t} = -\frac{1}{2}C_D\rho S_{\text{eff}}[u\vec{a} + (\vec{e}_u \cdot \vec{a})\vec{u}], \tag{3.18}$$

where $\vec{e}_u = \frac{\vec{u}}{u}$ is the unit vector in the direction of the vector $\vec{u}$. The lift force is given by the expression

$$\vec{F}_L = \frac{1}{2}C_L\rho S_{\text{eff}}u^2\vec{e}_L, \tag{3.19}$$

where $\vec{e}_L = \vec{n} \times \vec{e}_u$ and $\vec{n}$ is the unit vector perpendicular to the plane of the skier's motion, shown in Figure 3.16. Differentiating with respect to time and using equation (3.17) we obtain

$$\frac{\mathrm{d}\vec{e}_u}{\mathrm{d}t} = -\frac{1}{u}(\vec{e}_u \cdot \vec{a})\vec{e}_u + \frac{1}{u}\vec{a}, \tag{3.20}$$

which leads to

$$\frac{\mathrm{d}\vec{e}_L}{\mathrm{d}t} = \vec{n} \times \frac{\mathrm{d}\vec{e}_u}{\mathrm{d}t} = -\frac{1}{u}(\vec{e}_u \cdot \vec{a})\vec{e}_L + \frac{1}{u}\vec{n} \times \vec{a}. \tag{3.21}$$

Using equations (3.17), (3.19), and (3.21) we find

$$\frac{\mathrm{d}\vec{F}_L}{\mathrm{d}t} = \frac{1}{2}C_L\rho S_{\text{eff}}[u(\vec{e}_u \cdot \vec{a})\vec{e}_L + u(\vec{n} \times \vec{a})]. \tag{3.22}$$

It follows from equations (3.14), (3.18), and (3.22) that the jerk in initial moment is

$$\vec{j} = -\frac{C_D\rho S_{\text{eff}}}{2m}[u_0\vec{a}_0 + (\vec{e}_u \cdot \vec{a}_0)\vec{u}_0] + \frac{C_L\rho S_{\text{eff}}}{2m}[u_0(\vec{e}_u \cdot \vec{a}_0)\vec{e}_L + u_0(\vec{n} \times \vec{a}_0)], \tag{3.23}$$

where $\vec{u}_0 = \vec{v}_0 - \vec{w}$.

(f) When the jerk does not depend on time, using the solution of part (d), we find that the time dependence of the skier's coordinates reads

$$x(t) = v_{0x}t + \frac{1}{2}a_{0x}t^2 + \frac{1}{6}j_xt^3, \tag{3.24}$$

$$y(t) = v_{0y}t + \frac{1}{2}a_{0y}t^2 + \frac{1}{6}j_yt^3, \tag{3.25}$$

where v_{0x} and v_{0y} are the components of the skier's velocity immediately after takeoff, while a_{0x} and a_{0y} are the components of acceleration in that moment. In the moment of landing the coordinates satisfy the condition

$$\tan\delta = -\frac{y(t)}{x(t)}. \tag{3.26}$$

From equations (3.24)–(3.26) the moment of landing t satifies the equation

$$\frac{1}{6}(j_x\tan\delta + j_y)t^2 + \frac{1}{2}(a_{0x}\tan\delta + a_{0y})t + v_{0x}\tan\delta + v_{0y} = 0, \tag{3.27}$$

where

$$v_{0x} = v_0\cos\gamma_0, \tag{3.28}$$

$$v_{0y} = -v_0\sin\gamma_0. \tag{3.29}$$

Newton's second law for the motion of the skier implies

$$ma_{0x} = F_L\sin\gamma - F_D\cos\gamma, \tag{3.30}$$

$$ma_{0y} = -mg + F_D\sin\gamma + F_L\cos\gamma, \tag{3.31}$$

where the angle γ satisfies $\tan\gamma = -\frac{v_{0y}-w_y}{v_{0x}-w_x}$, which leads to

$$a_{0x} = \frac{1}{2}\frac{\rho S_{\text{eff}}}{m}(\vec{v}_0 - \vec{w})^2(C_L\sin\gamma - C_D\cos\gamma), \tag{3.32}$$

$$a_{0y} = -g + \frac{1}{2}\frac{\rho S_{\text{eff}}}{m}(\vec{v}_0 - \vec{w})^2(C_L\cos\gamma + C_D\sin\gamma). \tag{3.33}$$

The components of the jerk are given by the expressions

$$j_x = -\frac{C_D\rho S_{\text{eff}}}{2m}[u_0 a_{0x} + (\vec{e}_u\cdot\vec{a}_0)u_{0x}] + \frac{C_L\rho S_{\text{eff}}}{2m}[u_0(\vec{e}_u\cdot\vec{a}_0)\sin\gamma - u_0 a_{0y}], \tag{3.34}$$

$$j_y = -\frac{C_D\rho S_{\text{eff}}}{2m}[u_0 a_{0y} + (\vec{e}_u\cdot\vec{a}_0)u_{0y}] + \frac{C_L\rho S_{\text{eff}}}{2m}[u_0(\vec{e}_u\cdot\vec{a}_0)\cos\gamma + u_0 a_{0x}]. \tag{3.35}$$

When there is no wind, the equality $\vec{w} = 0$ holds. By solving the quadratic equation (3.27) with respect to t, where the coefficients in that equation are calculated using equations (3.28), (3.29), (3.32), (3.33), (3.34), and (3.35), we find

$$t = 3.709378\,\text{s} \approx 3.71\,\text{s}. \tag{3.36}$$

The smaller of the two solutions of the quadratic equation is chosen as physically relevant. The length of the jump is given by the expression

$$d = \frac{x(t)}{\cos \delta},$$ (3.37)

which leads to

$$d = 111.568679 \, \text{m} \approx 111.6 \, \text{m}.$$ (3.38)

(g) In the case of headwind, the components of the wind velocity are

$$w_x = -w \cos \delta,$$ (3.39)

$$w_y = w \sin \delta.$$ (3.40)

By solving quadratic equation (3.27) we obtain

$$t = 3.924928 \, \text{s} \approx 3.92 \, \text{s},$$ (3.41)

which leads to

$$d = 117.049816 \, \text{m} \approx 117.0 \, \text{m}.$$ (3.42)

In the case of tailwind, the components of the wind velocity are

$$w_x = w \cos \delta,$$ (3.43)

$$w_y = -w \sin \delta.$$ (3.44)

We find in this case

$$t = 3.537517 \, \text{s} \approx 3.54 \, \text{s}$$ (3.45)

and

$$d = 107.228986 \, \text{m} \approx 107.2 \, \text{m}.$$ (3.46)

These results imply that the headwind increases the length of the jump while the tailwind shortens the length of the jump. This result may seem counterintuitive, but one should bear in mind that the headwind lifts the skier, which helps in extending the length of the jump.

(h) From the conservation of energy, we find that the skier's velocity at takeoff is given by the expression $v_0 = \sqrt{2g(h + \Delta h)}$, where h is the difference in altitude determined in part (a). This leads to $v_0 = 25.195436 \, \frac{\text{m}}{\text{s}} \approx 25.2 \, \frac{\text{m}}{\text{s}}$. Since there is no wind, the equality $w = 0$ holds. By solving quadratic equation (3.27) we obtain in this case

$$t = 3.750950 \, \text{s} \approx 3.75 \, \text{s}$$ (3.47)

and

$$d = 113.642341\,\text{m} \approx 113.6\,\text{m}. \tag{3.48}$$

We refer the reader interested in more details about the physics of ski jumping to references [33] and [34].

Problem 18 Diving

How long can a diver use a diving cylinder?
Can a diver recognize where a sound comes from?
Why is a diving mask needed to see clearly?

In this problem we investigate several effects that occur when people are diving in the sea. Atmospheric pressure is $p_{\text{atm}} = 101.3\,\text{kPa}$, the density of seawater is $\rho = 1,020\,\frac{\text{kg}}{\text{m}^3}$, and gravitational acceleration is $g = 9.81\,\frac{\text{m}}{\text{s}^2}$.

Since there is no oxygen in water, a person needs to use a diving cylinder that contains a mixture of oxygen and other gases in order to stay in water for longer periods of time. A typical scheme of a diving cylinder is shown in Figure 3.17. It consists of a tank that contains the gas at a high pressure $p_1 = 15.0\,\text{MPa}$ (when the tank is filled) and a pressure regulator that reduces the pressure of the gas to the surrounding pressure p_h. Consequently the diver inhales the gas at pressure p_h.

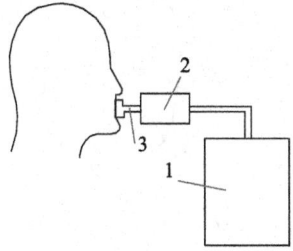

Figure 3.17 Scheme of a diving cylinder: (1) the gas at high pressure p_1, (2) pressure regulator, (3) the gas at a pressure p_h that the diver inhales

Although oxygen is necessary for breathing, inhalation of oxygen at a high partial pressure is harmful and leads to oxygen intoxication. The highest partial pressure for which it is safe to inhale oxygen is $p_{O_2}^{\text{max}} = 140\,\text{kPa}$.

(a) What is the maximal diving depth of a person who uses a diving cylinder that contains pure oxygen?

Since the depth obtained in part (a) is quite small, mixtures of gases are used for diving at larger depths. If nitrogen is used in the mixture of gases, its partial pressure must be smaller than $p_{N_2}^{\text{max}} = 400\,\text{kPa}$ since larger pressures lead to nitrogen narcosis.

(b) What is the maximal diving depth of a person who uses a diving cylinder that contains: (1) air with molar fraction of nitrogen of $r_{N_2}^v = 80\%$, and molar fraction of oxygen of $r_{O_2}^v = 20\%$; (2) the mixture known as Nitrox I that contains $r_{N_2}^n = 68\%$ of nitrogen and $r_{O_2}^n = 32\%$ of oxygen; (3) the mixture known as Trimix that contains $r_{N_2}^t = 20\%$ of nitrogen, $r_{O_2}^t = 10\%$ of oxygen and $r_{He}^t = 70\%$ of helium?

A diver at a depth of $h = 30.0\,\text{m}$ uses a diving cylinder with Trimix. The diver inhales oxygen of mass $m' = 83\,\frac{\text{g}}{\text{h}}$ in unit time. The volume of the tank at high pressure is $V_1 = 3.0\,\text{dm}^3$. The molar mass of oxygen is $M = 32.0\,\frac{\text{g}}{\text{mol}}$, while the gas constant is $R = 8.31\,\frac{\text{J}}{\text{mol}\cdot\text{K}}$. The whole system is at the same temperature $T = 300\,\text{K}$.

(c) How long can the diver use this diving cylinder?

We next consider the effects related to propagation of sound under water. In air, a person has the capability to estimate the direction from which a sound that is heard arrives. This is accomplished by registering the time interval between the arrival of the sound to the left ear and the right ear. This time interval is different in water due to the different speed of sound.

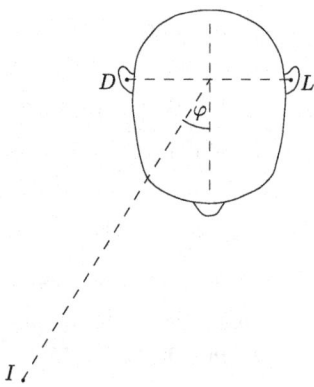

Figure 3.18 A person and the source of sound I. The right ear of the person is denoted as D, while the left ear is denoted as L.

(d) A person registers a sound emitted by a remote source I. The sound arrives from the direction which is at an angle φ with respect to the line perpendicular to the line that connects the ears; see Figure 3.18. The speed of sound in the medium is c_s, and the length of the line LD that connects the ears is l. Determine the time interval t_{DL} between when the sound arrives to the right ear and to the left ear.

(e) The person and the source of the sound are on the ground in air. If $t_{DL} =$ 0.416 ms, $l = 20.0$ cm, and the speed of sound in air is $c_a = 340\,\frac{m}{s}$, determine the direction from which the sound arrives if it is known that the source of sound is in front of the person.

(f) The person estimates the time interval t_{DL} with an accuracy of $\Delta t_{DL} = 0.07$ ms. How accurately will the person estimate the direction from which the sound arrives under the conditions of part (e)?

(g) The diver registers a sound that comes from a source that is in water. If the sound comes from the direction defined by the angle $\varphi = 45.0°$, calculate the time interval t_{DL} between when the sound arrives to the right ear and to the left ear of the diver. The speed of sound in water is $c_w = 1,500\,\frac{m}{s}$.

(h) The brain of the person determines the direction of the sound based on the time interval between when the sound arrives at the left and right ears, assuming that the person and the source of the sound are in air. The brain does not take into account that the speed of sound in water is different than in air. What direction will the diver estimate for the source of sound considered in part (g)?

We assumed in previous parts of the problem that the speed of sound is constant. In reality, the speed of sound depends on depth. In the range of depths of interest in this problem, this dependence can be approximated as $c(h) = c_0 - \alpha h$, where $c_0 = 1,500\,\frac{m}{s}$ and $\alpha = 0.0400\,\mathrm{s}^{-1}$. The diver is at a depth of $H = 100$ m, while a motorboat is slowly moving at the surface of water as shown in Figure 3.19.

(i) What is the largest possible distance between the diver and the boat that will still allow the diver to hear the sound of the boat? You can use justified approximations related to the fact that the change of speed of sound when the depth changes is small.

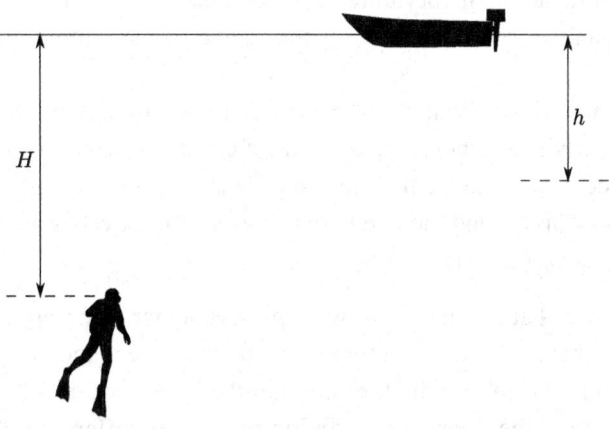

Figure 3.19 The diver and the motorboat

Interesting optical effects also occur underwater where the diver sees objects differently than in air.

(j) The diver can see light sources that are above the surface of water. Nevertheless, it occurs to the diver that these light sources exist only in the part of space determined by the condition that the angle between the light ray that arrives to the diver and the vertical is smaller than θ_{max}. Calculate the angle θ_{max}. The refractive index of water is $n_w = 1.33$.

The diver cannot clearly see objects in water unless a diving mask is used. To understand why this is the case, we consider the model of a human eye shown in Figure 3.20. Light rays that come from the surroundings refract at the cornea with radius of curvature R. The refractive index inside the eye is $n_o = 1.40$. In the case of a person with regular eyesight who is in air, the image of a remote object forms at the retina, which is located at a distance of $d_0 = 2.50$ cm from the cornea. Assume next that the angles between the relevant light rays and the optical axis of the system are small. The optical axis is defined by the direction that connects the object with the cornea and the retina (shown using a dashed line in Figure 3.20).

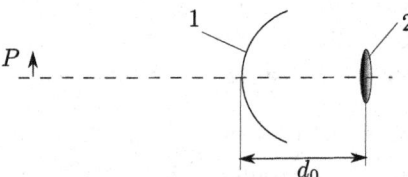

Figure 3.20 Simplified model of a human eye: (1) cornea, (2) retina. P is a remote object observed by the person.

(k) Determine the radius of curvature of the cornea.
(l) At what distance from the cornea is the image of the remote object when the person is in water?
(m) Determine the diopter value of the glasses the diver should wear in water in order to clearly see remote objects. Assume that the lenses of the glasses are at a distance of $d = 2.00$ cm from the cornea and that the space between the lenses of the glasses and the eye is filled with water. Lenses are made of glass whose refractive index is $n_s = 1.62$.

It is well known that divers do not wear glasses underwater, but they do wear a diving mask. In this case, there is a thin layer of air between the eyes and the water (see Figure 3.21). This allows the formation of the images of remote objects on the retina. Nevertheless, the diver with a diving mask sees differently than in air – it appears that the objects are larger and closer than they really are.

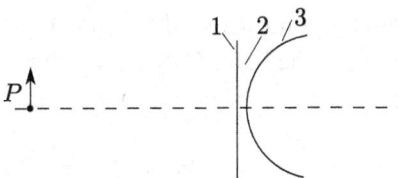

Figure 3.21 Scheme of the remote object P and the eye with the mask: (1) diving mask, (2) thin layer of air, (3) cornea

(n) How many times larger does the object appear to the diver?

(o) How many times closer does the object appear to the diver?

Solution of Problem 18

(a) The surrounding pressure at a depth h is $p_h = p_{atm} + \rho gh$. The condition $p_h < p_{O_2}^{max}$ has to be satisfied to avoid oxygen intoxication, which leads to $h < \dfrac{p_{O_2}^{max} - p_{atm}}{\rho g}$. Maximal depth at which one can dive with an oxygen diving cylinder is therefore $h_{max} = 3.87\,m$.

(b) Partial pressure of the oxygen at depth h is $p_{O_2} = r_{O_2}p_h$, while partial pressure of nitrogen is $p_{N_2} = r_{N_2}p_h$. The condition $p_{O_2} < p_{O_2}^{max}$ has to be satisfied to avoid oxygen intoxication, while the condition $p_{N_2} < p_{N_2}^{max}$ ensures that nitrogen narcosis does not happen. Maximal depth of diving is then determined by the condition

$$h_{max} = \min\left\{ \frac{\frac{p_{N_2}^{max}}{r_{N_2}} - p_{atm}}{\rho g}, \frac{\frac{p_{O_2}^{max}}{r_{O_2}} - p_{atm}}{\rho g} \right\}. \tag{3.49}$$

In the case of air the maximal depth is $h_{max}^V = 40\,m$, in the case of Nitrox I mixture it is $h_{max}^n = 34\,m$, and in the case of Trimix it is $h_{max}^t = 130\,m$.

(c) The diver will be able to breathe as long as the pressure in the tank at high pressure is larger than p_h. If the number of moles of the gas in the tank was n_m^1 at the beginning and n_m^2 at the end, the equations $p_1V_1 = n_m^1RT$ and $p_hV_1 = n_m^2RT$ hold. The number of moles of oxygen that the diver will inhale is $n_m = r_{O_2}^t \left(n_m^1 - n_m^2\right)$. On the other hand, the number of moles of oxygen inhaled in unit time is $n_m' = \frac{m'}{M}$. The time during which the diver will be able to use the cylinder is therefore $t = \frac{n_m}{n_m'}$. Using $p_h = p_{atm} + \rho gh$, we obtain from previous equations:

$$t = \frac{r_{O_2}^t \left(p_1 - p_{atm} - \rho gh\right)V_1M}{m'RT} = 41\,min. \tag{3.50}$$

(d) We denote as I the remote point where the source of sound is located, and we denote as C the midpoint of the line LD that connects the ears; see Figure 3.22. The time interval between the moment when the sound arrives to the right ear and to the left ear is

$$t_{DL} = \frac{LI - DI}{c_s}.$$ (3.51)

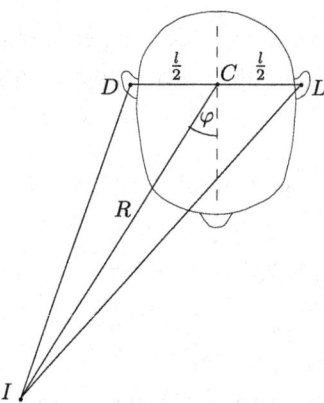

Figure 3.22 With the solution of problem 18(d)

The law of cosines applied to triangles LCI and DCI yields

$$LI = \sqrt{R^2 + \left(\frac{l}{2}\right)^2 - 2R\frac{l}{2}\cos\left(90° + \varphi\right)}$$ (3.52)

and

$$DI = \sqrt{R^2 + \left(\frac{l}{2}\right)^2 - 2R\frac{l}{2}\cos\left(90° - \varphi\right)},$$ (3.53)

where R is the length of the line CI. Starting from the equality

$$\sqrt{R^2 + \left(\frac{l}{2}\right)^2 - 2R\frac{l}{2}\cos\left(90° \pm \varphi\right)} = R\sqrt{1 + \left(\frac{l}{2R}\right)^2 - \frac{l}{R}\cos\left(90° \pm \varphi\right)}$$ (3.54)

and using the fact that $\frac{l}{R} \ll 1$ and the approximation $(1+x)^a \approx 1 + ax$ that holds for $x \ll 1$, it follows that

$$\sqrt{R^2 + \left(\frac{l}{2}\right)^2 - 2R\frac{l}{2}\cos\left(90° \pm \varphi\right)} \approx R - \frac{l}{2}\cos\left(90° \pm \varphi\right).$$ (3.55)

Using equations (3.51), (3.52), (3.53), and (3.55) we obtain

$$t_{DL} = \frac{l\sin\varphi}{c_s}. \tag{3.56}$$

(e) It follows from equation (3.56) that $\sin\varphi = \frac{c_s t_{DL}}{l}$, which leads to $\sin\varphi = 0.707$ — that is, $\varphi = 45.0°$ or $\varphi = 135.0°$. Since it is known that the source of sound is in front of the person, it follows that the sound arrived from the direction that is at an angle of $\varphi = 45.0°$ with respect to the line perpendicular to the line that connects the ears.

(f) Since the accuracy of determination of t_{DL} is $\Delta t_{DL} = 0.07$ ms, it follows that the person estimates that t_{DL} is from the interval (t_{min}, t_{max}), where $t_{min} = t_{DL} - \Delta t_{DL} = 0.346$ ms, and $t_{max} = t_{DL} + \Delta t_{DL} = 0.486$ ms. It follows that the person estimates that the angle φ is from the interval $(\varphi_{min}, \varphi_{max})$, where $\sin\varphi_{min} = \frac{c_s t_{min}}{l}$ and $\sin\varphi_{max} = \frac{c_s t_{max}}{l}$. It then follows that $\varphi_{min} = 36.0°$ and $\varphi_{max} = 55.7°$ — that is, $\varphi \approx (45° \pm 11°)$. The accuracy of determination of the direction is therefore $\Delta\varphi \approx 11°$.

(g) Using the solution of part (d), it follows that $t_{DL} = \frac{l\sin\varphi}{c_w} = 0.0943$ ms.

(h) The brain of the diver estimates the direction under the assumption that the speed of sound is c_a. Consequently the angle φ_p that defines the direction that the diver estimates satisfies the relation $\sin\varphi_p = \frac{c_a t_{DL}}{l}$, where t_{DL} is the time interval determined in part (g). This leads to $\varphi_p = 9.22°$. Consequently the diver makes a big error when estimating the direction of sound propagation in water and cannot determine this direction in water.

(i) We consider the propagation of the sound wave emitted by the engine of the motorboat. Its direction of propagation will change due to the change of speed of sound with the change of depth. We denote as $\beta(h)$ the angle between the direction of propagation and the vertical at the depth h, as shown in Figure 3.23. The law of refraction implies that the quantity $\frac{c(h)}{\sin\beta(h)}$ is invariant, which leads to

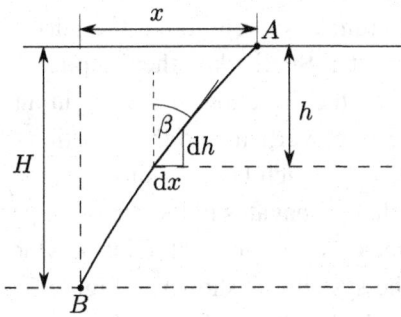

Figure 3.23 With the solution of part 18(i)

$$\frac{c(h)}{\sin \beta(h)} = \frac{c_0}{\sin \beta(0)}. \tag{3.57}$$

The change of the x-coordinate of the wave front when the depth changes by dh is $dx = \tan\beta(h) \cdot dh$ (see Figure 3.23). It follows that the difference of the x-coordinates of the boat and the diver is

$$x = \int_0^H dh \cdot \tan\beta(h). \tag{3.58}$$

The quantity x will be maximal when $\beta(0)$ is maximal – that is, when $\beta(0) \to 90°$. It then follows from equation (3.57) that

$$\sin\beta(h) = 1 - \frac{\alpha}{c_0}h, \tag{3.59}$$

while $\cos\beta(h) = \sqrt{1 - \sin^2\beta(h)}$. By replacing equation (3.59) in the previous expression, using the fact that $\frac{\alpha h}{c_0} \ll 1$ and the approximation $\left(1 - \frac{\alpha h}{c_0}\right)^2 \approx 1 - \frac{2\alpha h}{c_0}$, it follows that

$$\cos\beta(h) = \sqrt{\frac{2\alpha h}{c_0}}. \tag{3.60}$$

By replacing equations (3.59) and (3.60) in equation (3.58) and neglecting the small term $\frac{\alpha}{c_0}h$ in equation (3.59) we find that

$$x = \int_0^H dh \cdot \sqrt{\frac{c_0}{2\alpha h}}. \tag{3.61}$$

It then follows

$$x = \sqrt{\frac{2c_0 H}{\alpha}} = 2.74\,\text{km}. \tag{3.62}$$

The largest distance between the boat and the diver that still allows the diver to hear the boat is then $d_{\max} = \sqrt{H^2 + x^2} = 2.74\,\text{km}$.

(j) A light ray emitted from a source above the surface of the water refracts at the surface of the water. Snell's law then implies $\sin\theta_a = n_w \sin\theta_w$, where θ_a is the angle between the direction of the ray in air and the vertical, while θ_w is the angle between the vertical and the direction of the ray in water. θ_w is maximal when $\sin\theta_a = 1$, which leads to $\sin\theta_{\max} = \frac{1}{n_w}$ – that is, $\theta_w = 48.8°$.

(k) We consider the ray that propagates in the direction parallel to the optical axis of the system and intersects the cornea at point O, which is at a distance h from the optical axis of the system, as shown in Figure 3.24. We denote as C the center of the sphere, which is the surface of the cornea near the optical axis of the system, as O' the projection of point O to the optical axis of the system, and as α the angle between the line CO and the optical axis of the system. The ray considered refracts at point O and intersects the optical axis at point F. We denote as β the angle between the refracted ray and the line CO. Snell's law implies

$$\sin \alpha = n_o \sin \beta. \tag{3.63}$$

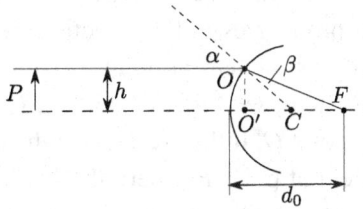

Figure 3.24 With the solution of problem 18(k)

By considering the right triangles $CO'O$ and $FO'O$ we, respectively, obtain the relations

$$\sin \alpha = \frac{h}{R}, \tag{3.64}$$

where R is the radius of curvature of the cornea and

$$\tan (\alpha - \beta) = \frac{h}{d_0}. \tag{3.65}$$

We next take into account that the angles α and β are small and we use the approximations $\tan \varphi \approx \sin \varphi \approx \varphi$, which are valid for small angles. From equations (3.63)–(3.65) we then obtain

$$R = d_0 \frac{n_o - 1}{n_o} = 0.714 \, \text{cm}. \tag{3.66}$$

(l) Using the same notation as in part (k), Snell's law for the ray that propagates in the direction parallel to the optical axis of the system reads

$$n_w \sin \alpha = n_o \sin \beta. \tag{3.67}$$

From the same considerations as in part (k) we obtain the relations

$$\sin \alpha = \frac{h}{R} \tag{3.68}$$

and

$$\tan (\alpha - \beta) = \frac{h}{d_w}, \tag{3.69}$$

where d_w is the distance from the cornea where the image of the remote object forms. Using the approximations for small angles, we obtain from equations (3.67)–(3.69) that

$$d_w = \frac{R n_o}{n_o - n_w} = 14.3 \,\text{cm}. \tag{3.70}$$

Since d_w is significantly different than d_0 the image does not form on the retina. Therefore, a human cannot clearly see in water without additional aids.

(m) We consider a ray that propagates in the direction parallel to the optical axis of the system and intersects the lens at point A, which is at a distance h from the optical axis of the system; see Figure 3.25. The ray refracts at the lens and intersects the cornea at point O. It then refracts at the cornea and intersects the optical axis of the system at point F where the image is formed. The person clearly sees the remote object when the point F is at the retina. Consequently the distance between point F and the cornea is d_0.

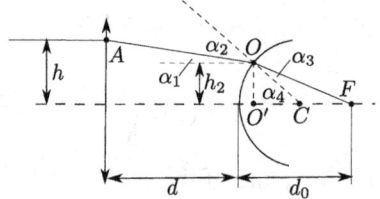

Figure 3.25 With the solution of part 18(m)

We denote as α_1 the angle between the line AO and the optical axis of the system. We then have

$$\tan\alpha_1 = \frac{h}{f_w}, \tag{3.71}$$

where f_w is the focal length of the lens in water, which is related to the focal length of the same lens in air f_a as

$$\frac{f_a}{f_w} = \frac{\frac{n_s}{n_w} - 1}{n_s - 1}. \tag{3.72}$$

We denote as O' the projection of the point O to the optical axis of the system and as h_2 the length of the line OO'. We then have

$$h_2 = h - d \cdot \tan\alpha_1. \tag{3.73}$$

It follows from the right triangle $CO'O$ that

$$h_2 = R\sin\alpha_4, \tag{3.74}$$

where $\alpha_4 = \angle OCO'$. Snell's law at point O gives

$$n_w \sin \alpha_2 = n_o \sin \alpha_3, \tag{3.75}$$

where $\alpha_3 = \angle FOC$, and α_2 is the angle between the ray AO and the line CO. By considering the right triangle $FO'O$ we obtain

$$\tan(\alpha_4 - \alpha_3) = \frac{h_2}{d_0}. \tag{3.76}$$

From the equality of angles with parallel rays, it also follows that

$$\alpha_4 = \alpha_1 + \alpha_2. \tag{3.77}$$

Next we take into account that all the angles considered are small and use the approximations valid for small angles. From equations (3.71)–(3.77) we then obtain

$$f_a = \left(d + \frac{d_0 R n_w}{d_0 n_w + n_o (R - d_0)}\right) \frac{n_s - n_w}{n_w (n_s - 1)} = 1.72 \, \text{cm} \tag{3.78}$$

– that is, the diopter value of these glasses should be $D_a = \frac{1}{f_a} = 58 \, \text{m}^{-1}$.

(n) We consider the ray that starts from point A at a remote object, refracts at the cornea in point O on the optical axis, and intersects the retina at point B, as shown in Figure 3.26. The influence of a thin layer of air on the trajectory of this ray can be neglected.

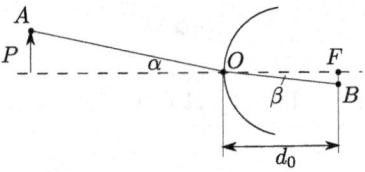

Figure 3.26 With the solution of part 18(n)

Snell's law gives

$$n_w \sin \alpha = n_o \sin \beta_w, \tag{3.79}$$

where α is the angle between the line AO and the optical axis, and β_w is the angle between the line BO and the optical axis. When the eye is in air, Snell's law for the same ray reads

$$\sin \alpha = n_o \sin \beta_a. \tag{3.80}$$

Using small angle approximations, it follows from equations (3.79) and (3.80) that

$$\beta_w = n_w \beta_a. \tag{3.81}$$

The size of the image on the retina is equal to the length of the line FB and reads $L = d_0 \tan\beta \approx d_0\beta$. It then follows that the size of the image on the retina is n_w times larger when the eye is in water. It follows that the objects that the person sees in water appear to be 1.33 times larger than in air.

(o) We consider a ray that departs from the object P and propagates at an angle α with respect to the optical axis of the system. The ray refracts at point Q at the boundary of water and air, and then it refracts again at point O on the cornea, as presented in Figure 3.27.

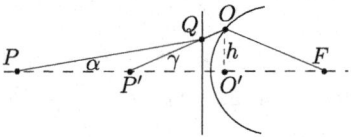

Figure 3.27 With the solution of part 18(o)

It appears to the human who is used to looking in air that the object is on the line OQ at the point P', which is the intersection of the line OQ with the optical axis. Points O and Q are approximately at the same position because the layer of air is thin. We denote as O' their projection to the optical axis and as h the length of the line OO'. It follows from the right triangle $PO'O$ that

$$\tan\alpha = \frac{h}{p}, \tag{3.82}$$

where p is the distance of the object from the cornea. Snell's law at point Q gives

$$n_w \sin\alpha = \sin\gamma, \tag{3.83}$$

where γ is the angle between the line OQ and the optical axis. The right triangle $P'O'O$ gives

$$\tan\gamma = \frac{h}{p'}, \tag{3.84}$$

where p' is the length of the line $P'O'$ and is equal to the apparent distance between the object and the cornea. From equations (3.82)–(3.84), using the small angle approximations, we find

$$p' = \frac{p}{n_w} \tag{3.85}$$

– that is, it appears to the diver that the object is 1.33 times closer.

Problem 19   Basketball

Is it easier for a taller or a shorter player to score?
Is the launching angle or the launching speed of the ball more important?

In a basketball game the players attempt to make shots that result in trajectories of the ball where the ball enters the basket. We consider such trajectories in this problem.

The basketball is moving in a homogeneous gravitational field of the Earth $\vec{g} = -g\vec{e}_y$, where $g = 9.81 \frac{m}{s^2}$. The trajectory of the ball is entirely determined by the velocity of the ball at time $t = 0$ when it leaves the hand of the player. We denote as v_0 the magnitude of this velocity and as α the angle between this velocity and the horizontal. Use the coordinate system defined in Figure 3.28 to describe the motion of the ball. Neglect air resistance.

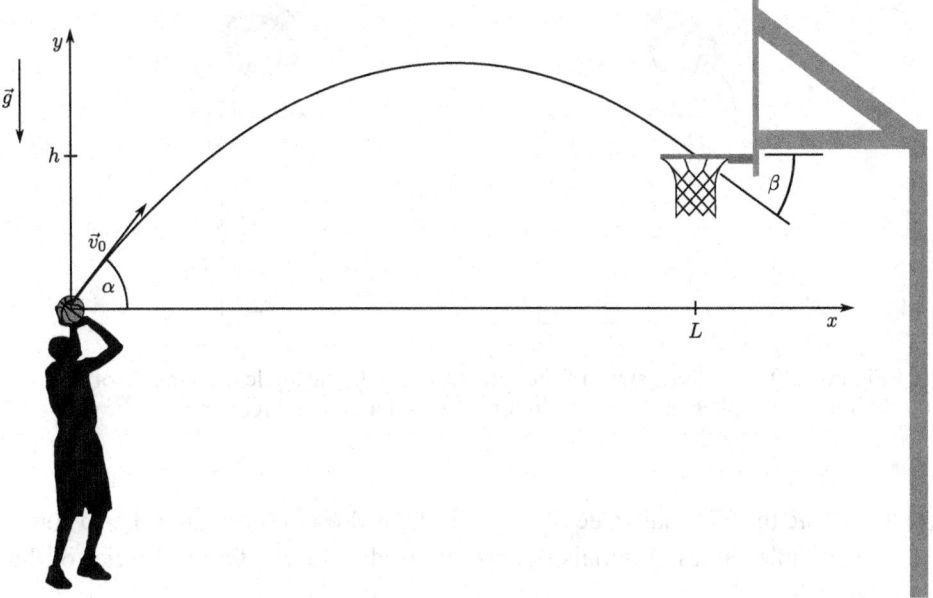

Figure 3.28   Coordinate system used to describe the motion of the ball

(a) Determine the components of the velocity of the ball and the coordinates of its position at time t. Show that the trajectory of the ball is a parabola.
(b) Find the angle between the velocity of the ball and the horizontal at the moment when the center of the ball is at the position (x,y). Express the tangent of this angle in terms of α, x, and y.

The position of the center of the basket rim is (L,h). We consider the trajectories of the ball that pass through this point. For the ball to pass through the basket, it must reach the rim from the top side, which means that the condition $\tan\alpha > h/L$ has to be satisfied.

(c) Determine the initial speed v_0 of a ball the trajectory of which passes through the center of the rim if the initial angle between its velocity and the horizontal is α.
(d) What is the launching angle α_0 for the minimal initial speed that enables the ball to pass through the center of the rim? What is the value of this minimal initial speed v_m?

The diameter of the rim is $D = 45.7\,\text{cm}$, while the diameter of the ball is $d = 23.8\,\text{cm}$. We denote as β the angle between the velocity of the ball and the horizontal at the moment when the ball enters the basket.

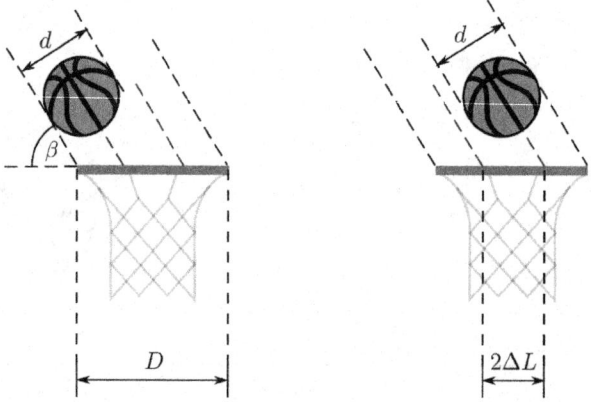

Figure 3.29 The dimensions of the rim and the ball, the angle of entrance of the ball into the basket, and the margin for error in horizontal direction ΔL

(e) Calculate the minimal value of angle β_0. How does the condition $\beta > \beta_0$ limit the possible values of initial angle α that lead to successful trajectories of the ball ?

When the condition $\beta > \beta_0$ is satisfied, the ball can pass through the basket even when it does not pass exactly through the center of the rim.

(f) Find the margin for error of the trajectory of the ball in horizontal direction ΔL in terms of d, D, and β that still allows the ball to pass through the basket (Figure 3.29).

We consider next the situation where the player scores by launching the ball at a speed determined in part (d) at an angle determined in that part of the problem. If the player launches the ball at a somewhat different speed (but still at the same angle), there is still a possibility that the player will score. In that case, the center of the ball will not pass exactly through the center of the rim.

(g) Calculate the range of launching speed that allows the player to score if $L = L_1 = 4.11$ m, $h = h_1 = 0.61$ m. Calculate this range if the player is significantly shorter so that $L_2 = 4.11$ m and $h_2 = 1.01$ m.

Next we assume that the player launches the ball at the speed determined in part (d) but that the launching angle is somewhat different than the angle determined in that part of the problem.

(h) Calculate the range of launching angles that allow the player to score if $L = L_1 = 4.11$ m, $h = h_1 = 0.61$ m.

Solution of Problem 19

(a) The vector of acceleration of the ball is $\vec{a} = -g\vec{e}_y$ and it has only the y-component. For this reason, the x-component of the velocity of the ball is constant and it is equal to its value at initial moment

$$v_x(t) = v_0 \cos \alpha. \tag{3.86}$$

The dependence of the x-coordinate of the ball on time is then

$$x(t) = v_0 t \cos \alpha. \tag{3.87}$$

To determine the y-coordinate and the y-component of the velocity, one can apply the formulas for motion with constant deceleration. One then obtains

$$v_y(t) = v_0 \sin \alpha - gt \tag{3.88}$$

and

$$y(t) = v_0 t \sin \alpha - \frac{1}{2}gt^2. \tag{3.89}$$

By eliminating the time from equations (3.87) and (3.89) we obtain

$$y = x \tan \alpha - \frac{g}{2v_0^2 \cos^2 \alpha} x^2, \tag{3.90}$$

which is the equation of the parabola.

(b) We denote as θ the angle between the velocity of the ball and the horizontal. This angle satisfies

$$\tan \theta = \frac{v_y}{v_x}. \tag{3.91}$$

We find from equations (3.86), (3.87), (3.88), (3.89), and (3.91) that

$$\tan\theta = 2\frac{y}{x} - \tan\alpha. \tag{3.92}$$

(c) By replacing $x = L$ and $y = h$ in equation (3.90) we find that the speed in question satisfies

$$v_0^2 = \frac{gL}{2\cos^2\alpha(\tan\alpha - \frac{h}{L})}. \tag{3.93}$$

(d) The speed v_0 will be minimal for the value of angle α when the function in the denominator of equation (3.93)

$$f(\alpha) = \cos\alpha(\sin\alpha - \frac{h}{L}\cos\alpha) \tag{3.94}$$

is maximal. The first derivative of this function is

$$f'(\alpha) = \cos(2\alpha) + \frac{h}{L}\sin(2\alpha). \tag{3.95}$$

It is equal to zero when

$$\tan(2\alpha_0) = -\frac{L}{h}, \tag{3.96}$$

while it is larger than zero when $\alpha < \alpha_0$ and smaller than zero when $\alpha > \alpha_0$. Consequently $f(\alpha)$ is maximal when $\alpha = \alpha_0$. We next find from equation (3.96) that

$$\alpha_0 = \frac{\pi}{4} + \frac{1}{2}\text{arctg}\frac{h}{L}. \tag{3.97}$$

By using trigonometric identities $\sin(2\alpha) = 2\sin\alpha\cos\alpha$ and $\cos^2\alpha = \frac{1+\cos(2\alpha)}{2}$ we can express the function $f(\alpha)$ as

$$f(\alpha_0) = \frac{1}{2}\sin(2\alpha_0) - \frac{h}{2L}[1 + \cos(2\alpha_0)]. \tag{3.98}$$

We find from equation (3.97) that

$$\sin(2\alpha_0) = \frac{1}{\sqrt{1 + \frac{h^2}{L^2}}}, \tag{3.99}$$

$$\cos(2\alpha_0) = -\frac{h}{L\sqrt{1 + \frac{h^2}{L^2}}}. \tag{3.100}$$

Since the minimal speed in question v_m satisfies the equation

$$v_m^2 = \frac{gL}{2f(\alpha_0)}, \tag{3.101}$$

we find from equations (3.98)–(3.101) that

$$v_m^2 = g\left(h + \sqrt{h^2 + L^2}\right). \tag{3.102}$$

(e) The case when the ball passes through the basket at the minimal possible angle β_0 is shown in Figure 3.30. When $\beta < \beta_0$ the ball will hit the front part of the rim and will not pass through the basket. We find from the figure that

$$\sin \beta_0 = \frac{d}{D} \Rightarrow \beta_0 = 31.4°. \tag{3.103}$$

The equation (3.92) describes the dependence of the angle between the velocity of the ball and the horizontal on the position of the ball. When the ball passes through the basket at a minimal angle, the relation $\theta = -\beta_0$ holds (the minus sign comes because the ball is moving downward at that moment), as well as $x = L$ and $y = h$. Consequently the minimal value of angle α satisfies the equation

$$\tan \alpha_{min} = \tan \beta_0 + \frac{2h}{L} = \frac{d}{\sqrt{D^2 - d^2}} + \frac{2h}{L}. \tag{3.104}$$

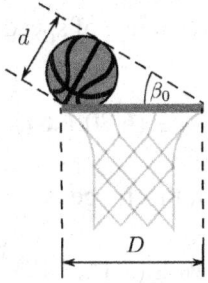

Figure 3.30 Passing of the ball through the basket at a minimal angle β_0

(f) We denote as B the center of the ball and as C_1 the center of the rim (Figure 3.31). Two limiting cases when the ball can still pass through the basket although its center does not pass through the center of the rim are shown in Figure 3.31. The distance BC_1 is equal to the margin for error ΔL. The triangle $\triangle ACB$ is right-angled with $BC = \frac{1}{2}d$, $AB = \frac{D}{2} - \Delta L$ and $\angle BAC = \beta$. Since $\sin \beta = \frac{BC}{AB}$, we find

$$\Delta L = \frac{1}{2}\left(D - \frac{d}{\sin \beta}\right). \tag{3.105}$$

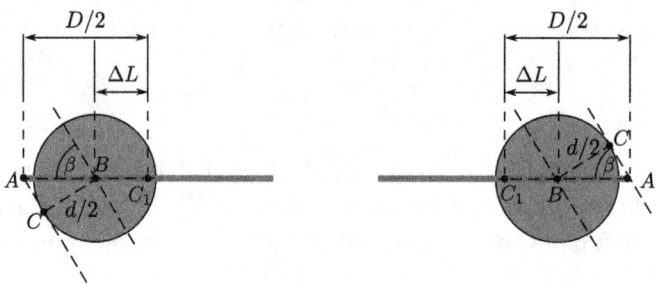

Figure 3.31 Passing of the ball through the basket at an angle $\beta > \beta_0$

(g) The player can score if the x-coordinate at the moment of passing of the ball
through the basket is in the range $(L - \Delta L, L + \Delta L)$, where ΔL is given by equation (3.105), while the angle β is determined from the relation [that comes
directly from equation (3.92)]

$$\tan\beta = \tan\alpha - \frac{2h}{L}, \tag{3.106}$$

and the angle α is determined by equation (3.97). The minimal and maximal
speed for which the player will still score are obtained by replacing $L - \Delta L$ and
$L + \Delta L$ in the expression for speed given by equation (3.93). After performing
the calculation, we find that the range of speed in question is in this case equal
to

$$v_0 \in (6.80, 6.87) \, \frac{m}{s}. \tag{3.107}$$

In the case of a shorter player, this range is

$$v_0 \in (7.15, 7.19) \, \frac{m}{s}. \tag{3.108}$$

We find that this interval is smaller for a shorter player. This means that the
margin of error is smaller for a shorter player and it is more difficult for this
player to score.

(h) Since we investigate the range of angles that allow the player to score, we first
transform equation (3.93) so that it is only the angle that appears at its left-hand
side. Using the trigonometric identity

$$\frac{1}{\cos^2\alpha} = 1 + \tan^2\alpha, \tag{3.109}$$

equation (3.93) becomes a quadratic equation with respect to $\tan\alpha$. Its solutions
are

$$\tan\alpha = \frac{v_0^2}{gL} \pm \sqrt{\left(\frac{v_0^2}{gL}\right)^2 - 1 - \frac{2v_0^2 h}{gL^2}}.$$ (3.110)

When $L = L_1$, both solutions of this equation are $\alpha = 49.2°$; the equation does not have real solutions when $L > L_1$, while for $L_1 - \Delta L \leq L < L_1$ the equation has two solutions that are equal to $\alpha = 45.6°$ and $\alpha = 52.9°$ when $L = L_1 - \Delta L$. This implies that for this value of initial speed, the trajectory will always end up at the point $x < L_1$. The player then scores when $L_1 - \Delta L \leq x < L_1$, which happens for

$$\alpha \in (45.6°, 52.9°).$$ (3.111)

We therefore see that the relative margin of error for the angle is much larger than the relative margin of error for the speed.

You can find more on the connection of physics and basketball in reference [4]. The authors thank Duško Latas for the initial version of this problem.

Problem 20   Football

How much time does a player have to react?
What happens during the bounce of the ball?
What is the probability of a given score during a game?

We consider in this problem the physics of football. Football is arguably the most popular sport in the world. Football matches are played between two teams on a grass field in the shape of a rectangle of size $100\,\text{m} \times 70\,\text{m}$. The circumference of the ball is $O = 70\,\text{cm}$, the mass of the ball is $m = 0.43\,\text{kg}$, and the pressure in the ball is $p = 0.85 \cdot 10^5\,\text{Pa}$ (more precisely, this is the difference between the pressure in the ball and the atmospheric pressure). There are $N = 11$ players on each team.

The ratio of the number of players in the field and the area of the field makes football a very dynamic game. This ratio was chosen so that something interesting is happening in each moment of the game.

(a) Assume that each player is assigned the same area in the field and that the players are moving at a speed of $v = 5.0\,\text{m/s}$. Estimate the mean time a player has to receive the ball, to control it, and to pass it to another player before the player of another team arrives.

We consider several phenomena related to the motion of the football. Consider a ball that falls in a vertical direction and bounces off the ground. The ball is not deformed before it reaches the ground and the pressure forces that act on the inner surface of the ball are in balance so that the total force that acts on the ball is zero. You can neglect gravity in this problem.

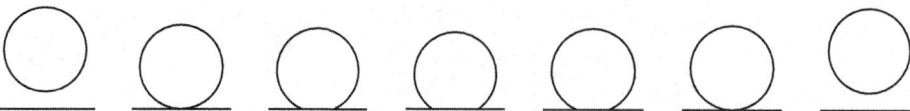

Figure 3.32 Bouncing of a ball recorded using a high-speed camera

When the ball touches the ground, it starts to deform, as shown in Figure 3.32. The deformation leads to the appearance of the resulting force on the ball, which leads to the bounce of the ball. The geometry of the deformation of the ball is shown in Figure 3.33. The speed of the ball in the moment when it starts to touch the ground is v_0.

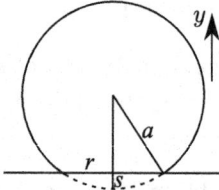

Figure 3.33 Geometry of the deformation of the ball during the bounce

The radius of the ball is a, $s \ll a$ denotes the size of the deformation, and r denotes the radius of circular area of the contact between the ball and the ground. Assume that the pressure inside the ball is constant, while the y-component of the velocity of the ball's center of mass is connected to the size of the deformation as

$$v_y = -\frac{ds}{dt}.$$

(b) Find the time dependence of the magnitude of the resulting force on the ball and determine its maximal value. Express your result in terms of the quantities O, p, m, and v_0.

(c) Calculate the duration of the contact of the ball with the ground.

When the ball bounces off the ground, part of its kinetic energy is lost (this effect was not considered in previous parts of the problem). The quantity describing this loss is the restitution coefficient $e = v_{after}/v_{before}$, where v_{after} and v_{before} are magnitudes of the vertical component of the velocity of the ball after and before the bounce. The restitution coefficient of grass is $e = 0.60$.

Consider a ball that hits the ground in such a way that the projections of the velocity on the horizontal and the vertical before the bounce are respectively u_1 and v_1, while the angular velocity is ω_1. The values of these quantities immediately after the impact are denoted using index 2. The directions of all velocities and angular velocities are given in Figure 3.34.

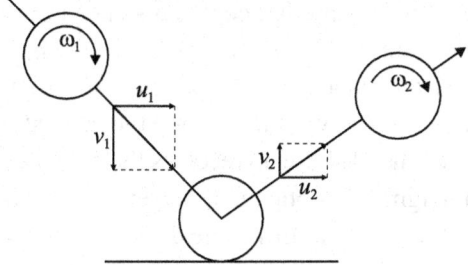

Figure 3.34 Bounce of the ball

(d) Determine the expression for projections of the velocity on the horizontal and the vertical and the angular velocity immediately after the bounce if the ball is sliding on the ground during the bounce. The sliding friction coefficient is $\mu = 0.70$. The moment of inertia of the ball with respect to the axis that passes through the center of the ball is $I = \frac{2}{3}ma^2$. Assume that $u_1 > a\omega_1$.

(e) When the ball slides during the whole duration of the bounce, the condition $u_2 > a\omega_2$ holds. If this condition is not satisfied, there will be no sliding at the end of the bounce and the ball will roll. Determine the range of incident angles of the ball that does not rotate that leads to sliding of the ball at the end of the bounce.

(f) Determine the expression for projections of the velocity on horizontal and vertical and angular velocity immediately after the bounce if the ball is rolling at the end of the bounce. In this part of the problem, assume that the ball also rotates before the bounce.

Inability to accurately predict the score of the football match is one of the things that makes football a very interesting sport. We assume that the efficiency of a football team can be described using one parameter r that describes the probability that the team scores a goal in a unit of time. The probability that the team scores n goals in time t is then given by the Poisson distribution

$$P_r(n,t) = \frac{(rt)^n}{n!}e^{-rt}.$$

(g) Determine the probability that the score at time t in the match of the teams with efficiencies r_1 and r_2, is $(n_1 : n_2)$.

(h) What is the probability that team 1 scores the goal first up to time t?

Solution of Problem 20

(a) Each team has N players and the area of the field is A, which implies that each player of one team controls the area $A_1 = \frac{A}{N}$. If we assume that the players of one team are positioned on a square lattice, the distance between two neighboring players of the same team will be $d_1 = \sqrt{A_1}$. Since the ball is usually passed to the player who is at a distance from opponents, we assume that the distance between the player who receives the ball and the nearest opponent is $d = \frac{1}{2}d_1$. This is satisfied when the player who receives the ball is at the center of the line that connects two neighboring opponent players. If the speed of the players is v, the time it takes the player to travel the distance d is $t = d/v$. Consequently the mean time it takes a football player to receive, control, and pass the ball is

$$t = \frac{1}{2v}\sqrt{\frac{A}{N}} = 2.5\,\text{s.} \tag{3.112}$$

(b) The forces that act on the part of the ball that is in contact with the ground are shown in the left part of Figure 3.35, where p_{in} is the pressure inside the ball, S_c the area of that part of the ball, while N is the reaction force of the ground. Since this part of the ball is at rest, it holds that $N = p_{in}S_c$. Vertical components of resulting external forces that act on the ball are shown in the right part of Figure 3.35. Vertical components of atmospheric pressure forces that act on part 1 of the ball cancel each other due to symmetry. The same holds for part 2 of the ball. Consequently the vertical component of the total resulting force that originates from external pressure p_{out} is $p_{out}S_c$, where S_c is the area of projection of the surface S_g to the horizontal plane. This area is exactly equal to the area of the part of the ball that is in contact with the ground. Newton's second law for the motion of the ball then reads $m\frac{dv_y}{dt} = N - p_{out}S_c$.

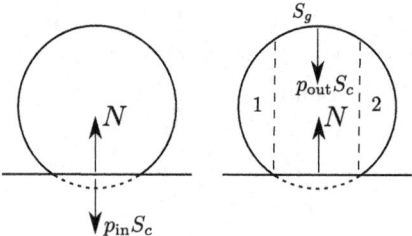

Figure 3.35 Forces that act on the football during the bounce

The geometry of the deformation is presented in Figure 3.33. We see from the figure that $r^2 = a^2 - (a-s)^2 = 2as - s^2$. Since $s \ll a$, the last term can be neglected. Consequently, the area of contact between the ball and the ground is $S_c = \pi r^2 = 2\pi as = Os$. v_y is related to s as $v_y = -\frac{ds}{dt}$ during the contact. Bearing in mind that $p = p_{in} - p_{out}$, it follows from previous equations that

$$m\frac{d^2s}{dt^2} = -Ops.$$ (3.113)

The solution of this differential equation is $s(t) = A\sin(\omega t) + B\cos(\omega t)$, where $\omega = \sqrt{Op/m}$, while A and B are integration constants that can be determined from initial conditions $s(t=0) = 0$ and $v_y(t=0) = -v_0$. We then obtain

$$s = \frac{v_0}{\sqrt{Op/m}}\sin\left(\sqrt{\frac{Op}{m}}t\right),$$ (3.114)

which implies that the magnitude of the resulting force $F = \left|m\frac{d^2s}{dt^2}\right|$ changes as

$$F = v_0\sqrt{Opm}\sin\left(\sqrt{\frac{Op}{m}}t\right),$$ (3.115)

while its maximum value is

$$F_{\text{max}} = v_0\sqrt{Opm}.$$ (3.116)

(c) The time of the duration of the deformation is equal to the time interval between the start of the bounce and the moment when s becomes equal to zero again. From the condition $s(\tau) = 0$, using equation (3.114) it follows that the duration of the deformation is

$$\tau = \pi\sqrt{\frac{m}{Op}} = 8.4 \cdot 10^{-3}\,\text{s}.$$

(d) The vertical component of the velocity of the ball after the bounce is determined from the value of the restitution coefficient

$$v_2 = ev_1.$$ (3.117)

Since it was stated in the problem that $u_1 > a\omega_1$ and the ball slides on the ground during the collision, it follows that the direction of the friction force F_{tr} is opposite to the direction of the velocity $\vec{u}_1$. Newton's second law for motion in horizontal direction then gives

$$m(u_2 - u_1) = -\int_{t_-}^{t_+} F_{\text{tr}}\,dt,$$ (3.118)

where the times t_- and t_+ denote the moments immediately before and immediately after the collision. Newton's second law for motion in vertical direction yields

$$m(v_2 + v_1) = \int_{t_-}^{t_+} N\,dt.$$ (3.119)

Since the ball slides on the ground, it holds that

$$F_{tr} = \mu N. \tag{3.120}$$

Newton's second law for rotation implies

$$I(\omega_2 - \omega_1) = \int_{t_-}^{t_+} F_{tr} a \, dt, \tag{3.121}$$

where $I = \frac{2}{3}ma^2$ is the moment of inertia. We find from equations (3.117)–(3.121) that

$$v_2 = ev_1, \tag{3.122}$$

$$u_2 = u_1 - \mu(1+e)v_1, \tag{3.123}$$

$$\omega_2 = \omega_1 + \frac{3}{2}\mu(1+e)\frac{v_1}{a}. \tag{3.124}$$

(e) The condition $u_2/\omega_2 a \leq 1$ has to be satisfied for the ball to roll at the end of the collision. Using equations (3.123) and (3.124) this gives

$$\mu(1+e)v_1 \geq \frac{2}{5}(u_1 - \omega_1 a). \tag{3.125}$$

In the case of the ball that does not rotate before the bounce ($\omega_1 = 0$), the last inequality, using $\tan\theta = \frac{v_1}{u_1}$, becomes

$$\tan\theta \geq \frac{2}{5\mu(1+e)}, \tag{3.126}$$

where θ is the angle between the direction of motion of the ball and the ground. In the case of numerical values given in the problem, we find

$$\theta > 20°.$$

(f) In this case, the ball will slide on the ground at the beginning of the collision. In certain moment of time, the sliding will stop. After that moment, the horizontal component of the velocity and angular velocity will be constant, while the friction force will be zero. Equations (3.117), (3.118) and (3.121) are still satisfied in this case. From these equations and the condition that there is no sliding at the end of the collision $u_2 = \omega_2 a$ we find

$$v_2 = ev_1, \quad u_2 = \frac{3}{5}u_1 + \frac{2}{5}\omega_1 a, \quad \omega_2 = \frac{2}{5}\omega_1 + \frac{3}{5}\frac{u_1}{a}.$$

(g) The probability that the score at time t is $(n_1 : n_2)$ is equal to the product of the probability that the first team scores n_1 goals in time t and the probability that the second time scores n_2 goals in time t (since these events are independent in our model)

$$P_{(n_1:n_2)} = \frac{(r_1 t)^{n_1} (r_2 t)^{n_2}}{n_1! n_2!} e^{-(r_1 + r_2)t}.$$

(h) The probability that at least one goal is scored in time t is $P_g = 1 - P_{(0:0)}$. Let P_1 be the probability that the first team scores the goal first in time t and P_2 the probability that the second team scores the goal first in time t. We then have $P_1 + P_2 = P_g$. On the other hand, the ratio of these probabilities is equal to the ratio of team efficiencies $P_1/P_2 = r_1/r_2$. The last three equations imply that $P_1 = \frac{r_1}{r_1 + r_2}(1 - P_{(0:0)})$, which leads to $P_1 = \frac{r_1}{r_1 + r_2}(1 - e^{-(r_1 + r_2)t})$.

You can find a lot more details about physics in football in reference [38]. This problem was given by Duško Latas at the federal physics competition for the fourth grade of high school in Serbia and Montenegro in 2003.

Problem 21   Tennis

How to hit a tennis ball to impart a high speed or a high spin on it.

Tennis is a sport that was first played in the nineteenth century in England. The players use a racket to hit the ball and aim to hit it in such a way that the opponent cannot return it. For this reason, the players aim at imparting either a high speed or a high angular velocity (spin) to the ball. In the first case, the opponent does not have enough time to return the ball, while in the second case, the ball bounces in an inconvenient manner and it is difficult for the opponent to hit it in a controlled way. In this problem we consider the conditions that need to be satisfied for the tennis player to impart a high speed or a high spin to the ball.

The mass of a tennis ball is $m = 57.0\,\text{g}$, the mass of a tennis racket is $M = 180\,\text{g}$, and the moment of inertia with respect to the axis o that is in the plane of the racket and passes through the end point D of the racket handle (Figure 3.36) is $I = 0.0570\,\text{kg} \cdot \text{m}^2$. Assume that all collisions between the racket and the ball are

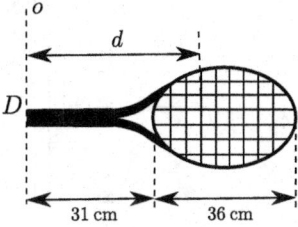

Figure 3.36 The scheme of a tennis racket

short so that the effect of gravity force can be neglected. In parts (a)–(d) assume that there are no energy losses – that is, that the initial kinetic energy is exchanged only between the racket and the ball.

We first consider how a tennis player should hit the ball when serving to impart it as high a speed as possible. The tennis player holds the racket at point D. Therefore, the racket can only rotate around axis o. Before the collision, the racket rotates around axis o at an angular velocity ω, while the ball is at rest and has no spin before the collision. The racket hits the ball at the point on the axis of the racket that is at distance d from point D (Figure 3.36).

(a) Express the speed of the ball v after the serve in terms of ω, d, I, and m.
(b) Determine by which part of the racket the ball should be hit so that its speed is maximal (minimal).

Block-return is a tennis shot in which a tennis player returns the opponent's serve just by placing the racket on the ball's trajectory. The ball is returned to the opponent this way. Assume that a ball with no spin that moves in the direction perpendicular to the racket hits the racket at the point on the racket axis that is at distance d from point D (Figure 3.36). The tennis player holds the racket at point D, so that the racket can only perform the rotation around axis o. The coefficient of restitution e is defined in this case as the ratio of the speed of the ball after and before the collision.

(c) Express e in terms of I, m, and d.
(d) Determine by which part of the racket the ball should be hit in block-return so that its speed is maximal (minimal).

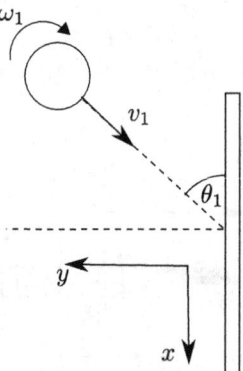

Figure 3.37 The collision of the ball with the racket

We next consider the collision of the ball with the racket in the case when the ball has the initial spin (Figure 3.37). We would like to understand what the ball's spin will be after the collision. The motion of the racket before the collision is purely translational. For simplicity, we assume that the collision of the ball with the racket does not cause rotation of the racket – that is, that the ball hits the center of mass of the racket. We consider the phenomenon in the frame of reference S in which the racket is at rest before the collision. Assume also that the hand of the player does not act on the racket during the collision of the ball and the racket.

The collision of the ball and the racket is not completely elastic in reality. A certain part of the energy is transferred to the racket vibrations and the internal energy of the ball. These energy losses are taken into account through the restitution coefficient, which is in this case defined as $e = \frac{v_{2y} - V_{2y}}{-v_{1y}}$, where v_{1y} and v_{2y} are the y-components of the velocity of the ball before and after the collision, V_{2y} is the y-component of the racket's velocity after the collision (all these velocities are defined in reference frame S). The coordinate system is defined in Figure 3.37. The apparent coefficient of restitution is defined as $e_P = \frac{v_{2y}}{-v_{1y}}$.

(e) Express e_P in terms of e, m, and M. Calculate the value of e_P if $e = 0.900$.

Depending on the parameters of the system, sliding of the ball on the racket or rolling (without sliding) can occur during the collision. We first consider rolling of the ball on the racket. In this case, the force of rolling friction occurs of magnitude $F = \mu_R N$, where μ_R is the rolling friction coefficient and N is the reaction force of the racket. Due to deformation of the racket and the ball, the reaction force of the racket is not directed toward the center of the ball, but has torque $N \cdot a$ with respect to the center of the ball (Figure 3.38). In the course of rotation in the direction shown in Figure 3.38, the force F is directed as shown in the same figure. Assume that the magnitude of its torque with respect to the center of the ball is $F \cdot R$, where R is the radius of the ball. The moment of inertia of the ball with respect to the axis that passes through the center of the ball is $I_L = \frac{2}{3} m R^2$.

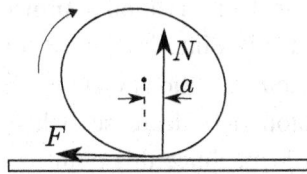

Figure 3.38 Rolling of the ball without sliding on the racket

(f) Express a in terms of R, μ_R, mass of the ball m, and mass of the racket M. Calculate the numerical value of a if $R = 32.5\,\text{mm}$ and $\mu_R = 0.0500$.

We next consider the case in which the ball is incident on the racket at an angle of $\theta_1 = 45.0°$ at a speed of $v_1 = 30.0\,\frac{m}{s}$ (in reference frame S, Figure 3.37). The angular velocity of the ball before the collision is $\omega_1 = 300\,\frac{rad}{s}$ and it is directed as in Figure 3.37. Neglect the torque of the force N during the sliding of the ball on the racket.

(g) Determine the dependence of the angular velocity of the ball after the collision ω_2 on the sliding friction coefficient μ_S between the racket and the ball.

(h) Plot the graph of the dependence of ω_2 on μ_S.

(i) For which value of μ_S will the ball change the direction of the spin during the collision and maximize the magnitude of the angular velocity?

Solution of Problem 21

(a) Conservation of energy yields $\frac{1}{2}I\omega^2 = \frac{1}{2}mv^2 + \frac{1}{2}I\omega_1^2$, where ω_1 is the angular velocity of the racket after the collision. The projection of angular momentum with respect to point D on axis o is also conserved, which leads to $I\omega = mvd + I\omega_1$. Previous equations lead to

$$v = \frac{2\omega Id}{I + md^2}. \qquad (3.127)$$

(b) Equation (3.127) leads to $v(d) = \frac{2\omega}{\frac{I}{d}+md}$. Therefore, the function $v(d)$ exhibits a maximum when the function $h(d) = \frac{I}{d} + md$ exhibits a minimum. Since $h'(d) = m - \frac{I}{d^2}$, it follows that $h'(d) < 0$ for $d < d_{gr} = \sqrt{\frac{I}{m}} = 1.00\,m$ – that is, $v(d)$ is monotonously increasing for $d < d_{gr}$. Since the point $d = d_{gr}$ is outside the racket, it follows that the ball has the highest speed if it is hit by the top of the racket – that is, the point with $d = 67\,cm$. The ball has the smallest speed if it is hit by the bottom of the racket – that is, the point with $d = 31\,cm$.

(c) Conservation of energy yields $\frac{1}{2}mv^2 = \frac{1}{2}mv_1^2 + \frac{1}{2}I\omega^2$, where v_1 is the speed of the ball after the collision with the racket, and ω is the angular velocity of the racket after the collision. Conservation of projection of angular momentum with respect to point D on axis o implies $mvd = I\omega - mv_1d$. The coefficient of restitution is $e = \frac{v_1}{v}$. Previous equations yield $e = \frac{I-md^2}{I+md^2}$.

(d) The function $e(d)$ monotonously decreases when d increases and takes the value 0 for $d = d_{gr} = 1.00\,m$. Since this point is outside the racket, the ball has the smallest speed when it is hit by the top of the racket ($d = 67\,cm$), and the largest if it is hit by the bottom of the racket ($d = 31\,cm$).

(e) Conservation of the y-component of the momentum implies $mv_{1y} = MV_{2y} + mv_{2y}$. Using the definitions of the quantities e and e_P, it follows that $e_P = \frac{eM-m}{M+m} = 0.443$.

(f) Newton's second law for translation of the ball and the racket gives $m\frac{dv_x}{dt} = -F$ and $M\frac{dV_x}{dt} = F$, while Newton's second law for rotation gives $I_L\frac{d\omega}{dt} = FR - Na$. The condition that there is no sliding is $v_x - R\omega = V_x$ — that is, $\frac{dv_x}{dt} - R\frac{d\omega}{dt} = \frac{dV_x}{dt}$. The equation $F = \mu_R N$ also holds. We find from previous equations that $a = R\mu_R\left(\frac{5}{3} + \frac{2m}{3M}\right) = 3.05$ mm.

(g) Two cases are possible depending on the value of μ_S: (1) the ball slides on the racket throughout the whole duration of the collision; (2) the ball slides at the beginning of the collision and then it rolls without sliding.

In the first case, the following equations hold: $I_L\frac{d\omega}{dt} = -FR$ and $m\frac{dv_y}{dt} = N$. By integrating from the moment before the collision t_1 to the moment after the collision t_2 we obtain

$$m\left(v_{2y} - v_{1y}\right) = \int_{t_1}^{t_2} N\,dt \tag{3.128}$$

and

$$I_L\left(\omega_2 - \omega_1\right) = -R\int_{t_1}^{t_2} F\,dt, \tag{3.129}$$

where the velocities and angular velocities with index 2 refer to moment t_2. Using $v_{2y} = -e_P v_{1y}$ and $F = \mu_S N$, and by eliminating the unknown integral from equations (3.128) and (3.129), we find

$$\omega_2 = \omega_1 - \frac{3v_1\sin\theta_1}{2R}\mu_S\left(1 + e_P\right). \tag{3.130}$$

In the second case, the following equations hold while the ball slides:

$$m\left(v_{0x} - v_{1x}\right) = -\int_{t_1}^{t_0} F\,dt, \tag{3.131}$$

$$I_L\left(\omega_0 - \omega_1\right) = -\int_{t_1}^{t_0} FR\,dt \tag{3.132}$$

and

$$MV_{0x} = \int_{t_1}^{t_0} F\,dt, \tag{3.133}$$

where t_0 is the moment when the ball stops sliding, while the quantities with the index 0 refer to that moment. From the condition for termination of sliding that reads $v_{0x} + R\omega_0 = V_{0x}$ and equations (3.131)–(3.133) we obtain

$$\int_{t_1}^{t_0} F\,dt = \frac{v_1\cos\theta_1 + R\omega_1}{\frac{1}{M} + \frac{5}{2m}}, \tag{3.134}$$

which implies $\omega_0 < 0$. By integrating Newton's second law throughout the whole duration of the collision (where the direction of the force F during sliding is determined, taking into account part [f] of the problem and the fact that $\omega_0 < 0$) we obtain the equations

$$I_L(\omega_2 - \omega_1) = -\int_{t_1}^{t_0} FR\,dt - \int_{t_0}^{t_2} FR\,dt + \int_{t_0}^{t_2} Na\,dt \qquad (3.135)$$

and

$$m(v_{2y} - v_{1y}) = \int_{t_1}^{t_0} N\,dt + \int_{t_0}^{t_2} N\,dt. \qquad (3.136)$$

Using equations (3.134)–(3.136), as well as the fact that $F = \mu_S N$ when the ball slides, and $F = \mu_R N$ when the ball rolls, we obtain

$$\omega_2 = \omega_1 - \frac{\omega_1 + \frac{v_1\cos\theta_1}{R}}{\frac{5}{3}+\frac{2m}{3M}} - (1+e_P)\frac{v_1\sin\theta_1}{R}\frac{3}{2}\left(\mu_R - \frac{a}{R}\right)$$

$$+ \frac{1}{\mu_S}\left(\mu_R - \frac{a}{R}\right)\frac{\omega_1 + \frac{v_1\cos\theta_1}{R}}{\frac{5}{3}+\frac{2m}{3M}}. \qquad (3.137)$$

Equalizing the expression for ω_2 in cases (1) and (2) we find that a borderline case between (1) and (2) occurs when $\mu_S^{\mathrm{gr}} = \frac{v_1\cos\theta_1 + R\omega_1}{(1+e_P)v_1\sin\theta_1\left(\frac{m}{M}+\frac{5}{2}\right)} = 0.359$. For $\mu_S < \mu_S^{\mathrm{gr}}$ the ball slides throughout the whole collision, while for $\mu_S > \mu_S^{\mathrm{gr}}$ sliding also occurs during the collision.

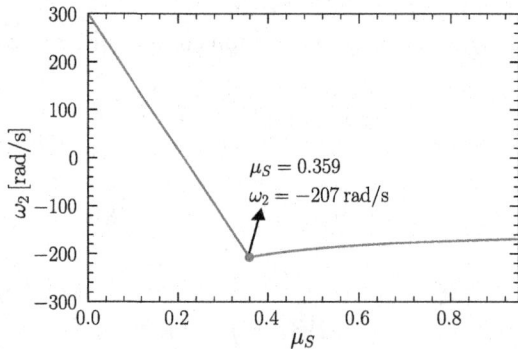

Figure 3.39 The graph of the dependence of ω_2 on μ_S

(h) The graph of the dependence of ω_2 on μ_S is shown in Figure 3.39.
(i) In the case when the spin of the ball changes direction during the collision, angular velocity after the collision has the highest magnitude when $\mu_S = \mu_S^{\mathrm{gr}}$.

We refer interested readers to references [7] and [8]. The authors presented this problem at the Serbian Physics Olympiad for high school students in 2013.

Problem 22   Cue Sports

How to hit a ball to maximize its speed after a collision?
How to hit a ball so that it returns to you?
What happens after a ball hits another, identical ball?

In this problem, we consider physical processes in cue sports – the motion of the ball, the collision of the cue and the ball, as well as the collision between the balls. Assume that a billiard ball is a homogeneous ball of radius R and mass m_b. The mass of the cue is m_s. The coefficient of sliding friction between the ball and the ground is μ_s. Unless otherwise stated, neglect the coefficient of rolling friction. Gravitational acceleration is g. Neglect air resistance.

(a) At some moment, the velocity of the ball was v_0, while its angular velocity was ω_0. Assumed directions of these quantities are shown in Figure 3.40. At that moment, the ball was sliding on the ground. Express the velocity of the ball in some later moment in terms of the angular velocity ω at that moment and the quantities v_0, ω_0, and R.

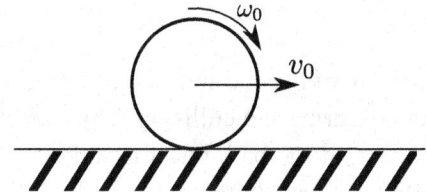

Figure 3.40 Billiard ball on horizontal surface

(b) Determine the velocity of the ball from part (a) in the moment when sliding stops. The expression should contain the quantities v_0, ω_0, and R.
(c) In the initial moment the angular momentum was imparted on the ball so that the magnitude of its angular velocity was ω_0 (where the vector of angular velocity is on the horizontal plane) and it was placed on a horizontal surface. The speed of translation was 0 in that moment. Assume in this part of the problem that the coefficient of rolling friction is μ_k. Determine the distance traveled by the ball before it stops. The expression should contain the quantities R, ω_0, g, μ_s, and μ_k.

We next consider the collision of the cue and the ball. In the following parts of the problem you can assume that the duration of this collision is very short.

The ball that was in rest on the horizontal surface was hit by the cue. The collision is absolutely elastic. In the moment of collision, the cue is in a horizontal position and the direction of the cue passes through the center of the ball. The speed of the cue in that moment is v_0, while the direction of this velocity coincides with the direction of the cue, as shown in Figure 3.41.

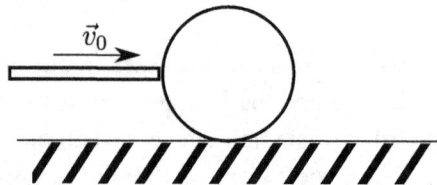

Figure 3.41 Collision of the cue and the ball

(d) Determine the speed of the ball immediately after the collision in terms of the quantities m_s, m_b, and v_0.

(e) Relative energy transfer during the collision of the cue and the ball is defined as the ratio of the kinetic energy of the ball immediately after the collision and the kinetic energy of the cue just before the collision. Express the relative energy transfer in terms of the quantity $\alpha = \frac{m_s}{m_b}$. For what α is the relative energy transfer maximal?

We next consider a collision of the cue and the ball that is not central. The direction of the cue is horizontal during the collision. This direction is on the vertical plane that contains the center of the ball, but it is shifted by b from the direction that contains the center of the ball, as shown in Figure 3.42. You can assume that the interaction force between the ball and the cue during their contact is constant and that its direction coincides with the direction of the cue. The collision is absolutely elastic.

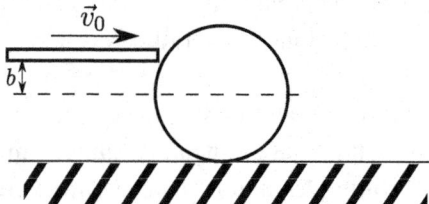

Figure 3.42 The collision of the cue and the ball that is not central

(f) Determine the ratio of the quantities $R\omega$ and v_b, where ω is the magnitude of angular velocity of the ball immediately after the collision, while v_b is the speed of the ball immediately after the collision. The expression should contain the quantities b and R.

(g) Determine the speed of the ball v_b immediately after the collision in terms of v_0, m_b, m_s, b, and R.

(h) Determine the speed of the ball v_k in the moment when sliding of the ball stops. The expression should contain the same quantities as in part (g).

(i) For which $\frac{b}{R}$ is v_k maximal? Assume that the condition $\frac{m_b}{m_s} < \frac{13}{2}$ is satisfied.

Under certain circumstances it is possible to hit the ball in such a way that it passes again through the initial position after some time. To achieve this, the ball that was at rest on the horizontal surface was hit by a cue at a point that is on the same horizontal plane as the center of the ball, as shown in Figure 3.43. The direction of the force by which the cue acts on the ball is the same as the direction of the cue, and it is at an angle of θ [$\theta \in (0°, 90°)$] with respect to horizontal. You can assume that the magnitude of this force during the collision is constant. The magnitude of the momentum of the force by which the cue acts on the ball is p (more precisely $p = F(t_+ - t_-)$, where F is the magnitude of the force, t_- is the moment just before the collision, and t_+ is the moment immediately after the collision).

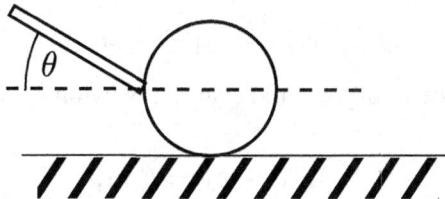

Figure 3.43 The collision of the cue and the ball in which it is possible that the ball returns to the initial position after some time

(j) What condition should the quantities θ and μ_s satisfy for the ball to pass again through the initial position after some time?

(k) For the case when the condition from part (j) is satisfied, determine the time interval after which the ball will pass again through the initial position. Express the result in terms of the quantities p, μ_s, m_b, g, and θ.

We finally consider the collision of two identical balls. There is no friction between the balls. A white ball whose speed is v that moves without sliding collides with a black ball of equal mass and radius, as shown in Figure 3.44. The collision is frontal – that is, the direction of the velocity coincides with the direction that connects the centers of the balls. Assume that the collision is absolutely elastic and that its duration is short – that is, that the duration t_s of the collision satisfies the condition $t_s \ll v/(\mu_s g)$.

(l) Determine the velocity and angular velocity of each of the balls immediately after the collision.

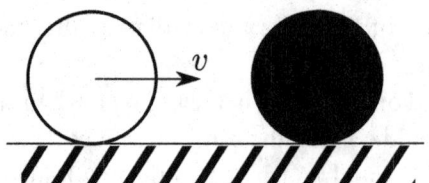

Figure 3.44 Collision of two balls

(m) Plot the graphs of the time dependence of the speed and angular velocity of each of the balls. Determine the expressions for all characteristic speeds, angular velocities, and times that appear on the graphs.

Solution of Problem 22

(a) Assumed direction of the velocity v, angular velocity ω, and the friction force F_{tr} are shown in Figure 3.45. Newton's second law for translation of the ball reads

$$m_b(v - v_0) = -F_{tr}t, \tag{3.138}$$

where v is the velocity of the ball at time t. Newton's second law for rotation yields

$$I(\omega - \omega_0) = F_{tr}Rt, \tag{3.139}$$

where ω is the angular velocity of the ball at time t, and $I = \frac{2}{5}m_b R^2$ is the moment of inertia of the ball with respect to the axis that passes through the center of the ball. The gravity force and the reaction force of the ground do not appear in previous equations because they cancel each other. We find from previous equations that

$$v = v_0 - \frac{2}{5}R(\omega - \omega_0). \tag{3.140}$$

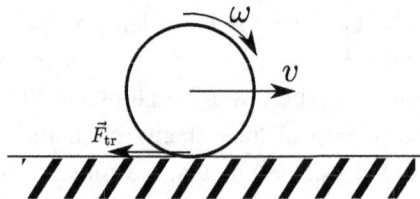

Figure 3.45 Assumed directions of the velocity, the angular velocity, and the friction force for the ball on horizontal surface

(b) In the moment when sliding stops, we have

$$v = R\omega.\tag{3.141}$$

Using equations (3.140) and (3.141) it follows that the velocity of the ball in the moment when sliding stops is

$$v = \frac{5v_0 + 2R\omega_0}{7}.\tag{3.142}$$

(c) After placing the ball on the ground, while it slides the sliding friction acts to decrease the magnitude of the angular velocity of the ball and to increase the speed of the ball. Using the solution of part (b), the speed of the ball in the moment when sliding stops is $v = \frac{2}{7}R\omega_0$. While the ball slides, the magnitude of friction force is $F_{\text{tr}}^{(1)} = \mu_s m_b g$ and therefore the acceleration of the ball is $a_1 = \frac{F_{\text{tr}}^{(1)}}{m_b} = \mu_s g$. The distance traveled by the ball while it slides is then $s_1 = \frac{v^2}{2a_1} = \frac{2}{49}\frac{(R\omega_0)^2}{\mu_s g}$. When sliding stops, the force of rolling friction acts on the ball. Its magnitude is $F_{\text{tr}}^{(2)} = \mu_k m_b g$ and it decelerates the ball. The magnitude of the deceleration is $a_2 = \frac{F_{\text{tr}}^{(2)}}{m_b} = \mu_k g$. The distance traveled by the ball from the moment when sliding stops to the moment when the ball stops is then $s_2 = \frac{v^2}{2a_2} = \frac{2}{49}\frac{(R\omega_0)^2}{\mu_k g}$. The total distance traveled by the ball before it stops is then $s = s_1 + s_2 = \frac{2(R\omega_0)^2}{49g}\left(\frac{1}{\mu_s} + \frac{1}{\mu_k}\right)$.

(d) We consider the physical system consisting of the cue and the ball. External forces acting on this system are the gravity force, the reaction of the ground acting on the ball, and the force of friction between the ball and the ground. Since the duration of the collision is short and all these forces have finite values, the change of the momentum of the system and the energy of the system during the collision is small. We can therefore apply the laws of conservation of energy and momentum to the moments just before and just after the collision. Conservation of horizontal component of momentum yields

$$m_s v_0 = m_s v_s + m_b v_b,\tag{3.143}$$

where v_s and v_b are horizontal components of the velocities of the cue and the ball just after the collision. Conservation of energy yields

$$\frac{1}{2}m_s v_0^2 = \frac{1}{2}m_s v_s^2 + \frac{1}{2}m_b v_b^2.\tag{3.144}$$

The system of equations (3.143) and (3.144) has two solutions: $v_s^{(1)} = v_0$, $v_b^{(1)} = 0$ and $v_s^{(2)} = \frac{m_s - m_b}{m_s + m_b}v_0$, $v_b^{(2)} = \frac{2m_s}{m_s + m_b}v_0$. The first solution corresponds to the moment before the collision, while the second corresponds to the moment after the collision. Therefore, the speed of the ball after the collision is

$$v_b = \frac{2m_s}{m_s + m_b} v_0.$$ (3.145)

(e) Relative transfer of energy during the collision is

$$r = \frac{\frac{1}{2}m_b v_b^2}{\frac{1}{2}m_s v_0^2}.$$ (3.146)

From equations (3.145) and (3.146) it follows that

$$r = \frac{4\alpha}{(1+\alpha)^2}.$$ (3.147)

Using the inequality $(1+\alpha)^2 \geq 4\alpha$, where equality holds when $\alpha = 1$, we find that maximal relative transfer of energy is obtained when $\alpha = 1$ and that it is equal to $r = 1$.

(f) The forces that act on the ball during the collision are: the interaction force between the ball and the cue whose magnitude is F, the gravity force, the force of reaction of the ground, and the friction force between the ball and the ground. The last three forces have finite value during the collision (the magnitude of the first two of them is $m_b g$, while the magnitude of the third is smaller than $\mu_s m_b g$). Since the duration of the collision is short, changes of the momentum and the angular momentum of the ball due to these three forces are negligible. On the other hand, the interaction force between the ball and the cue can have large values during the collision and corresponding changes of momentum and angular momentum cannot be neglected. Newton's second law for translation of the ball then yields

$$m_b v_b = F t,$$ (3.148)

where t is the duration of the collision and v_b is the horizontal component of the velocity of the ball just after the collision. Its assumed direction is shown in Figure 3.46. Newton's second law for rotation of the ball gives

$$I\omega = F b t,$$ (3.149)

where ω is the angular velocity of the ball just after the collision, while its assumed direction is shown in Figure 3.46. $I = \frac{2}{5}m_b R^2$ is the moment of inertia of the ball. From equations (3.148) and (3.149) we find

$$\frac{R\omega}{v_b} = \frac{5b}{2R}.$$ (3.150)

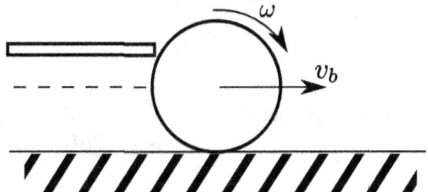

Figure 3.46 Assumed directions of the velocity and angular velocity of the ball after the collision of the ball and the cue that is not central

(g) For the same reasons as in part (d), conservation of momentum and energy in the moment just before and after the collision holds. Conservation of momentum yields

$$m_s v_0 = m_s v_s + m_b v_b, \tag{3.151}$$

where v_s is the horizontal component of the velocity of the cue just after the collision. Conservation of energy yields

$$\frac{1}{2}m_s v_0^2 = \frac{1}{2}m_s v_s^2 + \frac{1}{2}m_b v_b^2 + \frac{1}{2}I\omega^2. \tag{3.152}$$

From equations (3.150)–(3.152) it follows that

$$v_b = \frac{2v_0}{1 + \frac{m_b}{m_s} + \frac{5b^2}{2R^2}}. \tag{3.153}$$

(h) It follows from the solution of part (b) of the problem that the velocity of the ball in the moment when sliding stops is equal to $v_k = \frac{5}{7}v_b + \frac{2}{7}R\omega$. From the solution of part (f), it follows that $\frac{R\omega}{v_b} = \frac{5b}{2R}$. Using the solution of part (g) we find

$$v_k = \frac{10v_0}{7} \frac{1 + \frac{b}{R}}{1 + \frac{m_b}{m_s} + \frac{5b^2}{2R^2}}. \tag{3.154}$$

(i) v_k is maximal for the value of $x = \frac{b}{R}$ when the function

$$f(x) = \frac{1 + x}{1 + \frac{m_b}{m_s} + \frac{5}{2}x^2} \tag{3.155}$$

is maximal in the interval $x \in [-1, 1]$. The first derivative of this function is

$$f'(x) = \frac{-\frac{5}{2}x^2 - 5x + 1 + \frac{m_b}{m_s}}{\left(1 + \frac{m_b}{m_s} + \frac{5}{2}x^2\right)^2} \tag{3.156}$$

and it is equal to zero when

$$x_m^{(\pm)} = -1 \pm \sqrt{\frac{7}{5} + \frac{2m_b}{5m_s}}. \tag{3.157}$$

The solution $x_m^{(-)}$ is outside the interval $[-1, 1]$, while the solution $x_m^{(+)}$ is for $\frac{m_b}{m_s} < \frac{13}{2}$ in the interval $[-1, 1]$. For $x \in \left[x_m^{(+)}, 1 \right]$ it holds that $f'(x) < 0$, while for $x \in \left[-1, x_m^{(+)} \right]$ it holds that $f'(x) > 0$. Consequently the function $f(x)$ has a maximum at $x = x_m^{(+)}$. Therefore, v_k is maximal for $\frac{b}{R} = -1 + \sqrt{\frac{7}{5} + \frac{2m_b}{5m_s}}$.

(j) If the ball does not slide on the surface during the collision with the cue, it will continue to move without sliding after the collision. Its velocity will be constant and equal to the velocity obtained during the collision. Consequently it will not return to the initial position in this case. Therefore, we consider the case when the ball slides on the surface during the collision. From Newton's second law for translation in the x-direction, we find that the equation $(F \cos \theta - F_{tr})(t - t_-) = m_b v(t)$ holds during the collision, where $v(t)$ is the x-component of the velocity of the ball at time t during the collision, while F_{tr} is the friction force between the ball and the ground. From Newton's second law for rotation it follows that $(F \sin \theta - F_{tr}) R (t - t_-) = \frac{2}{5} m_b R^2 \omega(t)$, where $\omega(t)$ is the angular velocity of the ball at time t, with its assumed direction shown in Figure 3.47. From the condition that the ball does not move in the y-direction, it follows that $N = F \sin \theta + m_b g$, where N is the reaction force of the ground. When the ball slides, we have $F_{tr} = \mu_s N$. From previous equations we find

$$v(t) = (\cos \theta - \mu_s \sin \theta) \frac{F(t - t_-)}{m_b}, \tag{3.158}$$

$$R\omega(t) = \frac{5}{2} \sin \theta (1 - \mu_s) \frac{F(t - t_-)}{m_b}, \tag{3.159}$$

where all terms of the from $m_b g (t - t_-)$ were neglected because $(t - t_-)$ is small since the duration of the collision is short, and $m_b g$ is a finite quantity (the terms of the form $F(t - t_-)$ and $N(t - t_-)$ cannot be neglected because F and N do not necessarily have an upper limit). The ball slides during the collision when the condition $v(t) + R\omega(t) > 0$ is satisfied. Using equations (3.158) and (3.159) we then find $\mu_s < \frac{2}{7}\cot\theta + \frac{5}{7}$, which is the first necessary condition for the ball to return to its initial position.

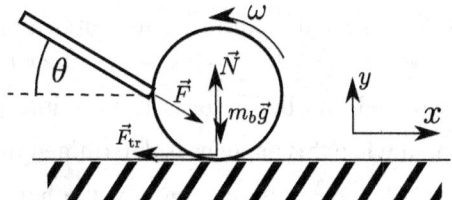

Figure 3.47 The collision of the cue and the ball in which it is possible that the ball returns to the initial position after some time. Assumed direction of the angular velocity and the forces acting on the ball are shown in the figure.

Next, by replacing $t = t_+$ in equations (3.158) and (3.159) we obtain the velocity and angular velocity of the ball just after the collision:

$$v_0 = (\cos\theta - \mu_s \sin\theta)\,\frac{p}{m_b}, \tag{3.160}$$

$$R\omega_0 = \frac{5}{2}\sin\theta\,(1 - \mu_s)\,\frac{p}{m_b}. \tag{3.161}$$

The following equations hold for the motion of the ball after the collision while the ball is sliding on the surface:

$$m_b\,[v(\tau) - v_0] = -F_{tr}\tau, \tag{3.162}$$

$$\frac{2}{5}m_b R^2\,[\omega(\tau) - \omega_0] = -F_{tr}R\tau \tag{3.163}$$

and

$$F_{tr} = \mu_s m_b g, \tag{3.164}$$

where τ is the time interval from the end of the collision, and $v(\tau)$ and $\omega(\tau)$ are the velocity and angular velocity in that moment. The ball stops sliding when the condition $v(\tau_k) + R\omega(\tau_k) = 0$ is satisfied. From this condition and equations (3.160)–(3.164) it follows that:

$$\tau_k = \frac{2p}{7\mu_s m_b g}\left(\cos\theta + \frac{5}{2}\sin\theta - \frac{7}{2}\mu_s\sin\theta\right). \tag{3.165}$$

The ball will return to its initial position if $v(\tau_k) < 0$. Using this condition and equations (3.160), (3.162), (3.164), and (3.165) it follows that $\theta > 45°$. **Therefore, the ball will return to its initial position if the conditions $\mu_s < \frac{2}{7}\cot\theta + \frac{5}{7}$ and $\theta > 45°$ are satisfied.**

(k) Since the deceleration of the ball after the end of the collision is $\mu_s g$, it will return to its initial position at time $\tau_v^{(1)} = \frac{2v_0}{\mu_s g}$, assuming that $\tau_v^{(1)} < \tau_k -$ that is, that sliding stops after the moment $\tau_v^{(1)}$. Using equations (3.160) and

(3.165) it follows that the ball returns to its initial position at time $\tau_v^{(1)} = \frac{2p}{\mu_s m_b g} (\cos \theta - \mu_s \sin \theta)$ if $\mu_s > \frac{12 \cot \theta - 5}{7}$. On the other hand, if the condition $\mu_s \leq \frac{12 \cot \theta - 5}{7}$ is satisfied, the ball returns to its initial position after sliding stops, and the moment when this happens is determined from $\tau_v^{(2)} = \tau_k + \frac{x(\tau_k)}{-v(\tau_k)}$, where $x(\tau_k) = v_0 \tau_k - \frac{1}{2} \mu_s g \tau_k^2$ is the position of the ball in the moment when sliding stops. Using equations (3.160), (3.162), (3.164), and (3.165) we find

$$\tau_v^{(2)} = \frac{2p}{35 \mu_s m_b g} \frac{\left(\cos \theta + \frac{5}{2} \sin \theta - \frac{7}{2} \mu_s \sin \theta\right)^2}{\sin \theta - \cos \theta}.$$

(l) The velocity of the white ball is, as stated in the problem, v. Since it does not slide, its angular velocity is $\omega = \frac{v}{R}$. The maximal momentum that can be imparted on the ball during the collision due to the action of the friction force is $p_s = \mu_s m_b g t_s$. From the condition stated in the problem $t_s \ll v/(\mu_s g)$, it follows that $p_s \ll m_b v$. Consequently, the change of the momentum of the ball due to the friction force during the collision can be neglected. In a similar manner we conclude that the changes of angular momentum and kinetic energy due to the friction force during the collision can be neglected. Momentum conservation in the moment just before and after the collision implies $m_b v = m_b v_1 + m_b v_2$, where v_1 and v_2 are the velocities of the white and black balls just after the collision. Since the change of angular momentum due to the friction force can be neglected and there is no friction between the balls, the angular velocities of the balls during the collision do not change. Conservation of energy just before and after the collision then yields $\frac{1}{2} m_b v^2 + \frac{1}{2} I \omega^2 = \frac{1}{2} m_b v_1^2 + \frac{1}{2} I \omega^2 + \frac{1}{2} m_b v_2^2$. The last two equations lead to $v_1 = 0$ and $v_2 = v$. It follows that the velocity of the translation of the white ball after the collision is $v_1 = 0$ while its angular velocity is $\omega_1 = \frac{v}{R}$. The velocity of the black ball in that moment is $v_2 = v$, while its angular velocity is $\omega_2 = 0$.

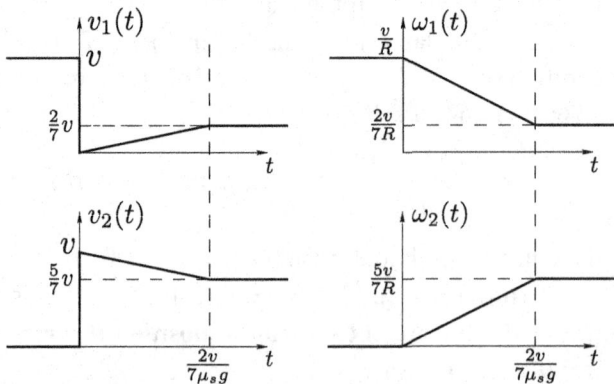

Figure 3.48 The graphs of the time dependence of the velocities and angular velocities of the balls

(m) The solution of part (l) implies that both balls slide on the surface just after the collision. For this reason, the sliding friction force acts on both of them. The magnitude of this force is $F = \mu_s m_b g$. Newton's second law for the translation and the rotation of the white ball gives $m_b a_1 = \mu_s m_b g$ and $I\alpha_1 = -\mu_s m_b g R$, while for the black ball $m_b a_2 = -\mu_s m_b g$ and $I\alpha_2 = \mu_s m_b g R$. The dependence of the velocities and angular velocities of the balls on time is then given as $v_1(t) = \mu_s g t$, $\omega_1(t) = \frac{v}{R} - \frac{5\mu_s g}{2R} t$, $v_2(t) = v - \mu_s g t$ and $\omega_2(t) = \frac{5\mu_s g}{2R} t$. This dependence holds until the moment when sliding stops. For the first ball it will happen at time t_1 that satisfies $v_1(t_1) = R\omega_1(t_1)$, while for the second ball this will happen at time t_2 that is obtained from $v_2(t_2) = R\omega_2(t_2)$. We find from these equations that $t_1 = \frac{2v}{7\mu_s g}$, $v_1(t_1) = \frac{2}{7}v$, $\omega_1(t_1) = \frac{2v}{7R}$, $t_2 = \frac{2v}{7\mu_s g}$, $v_2(t_2) = \frac{5}{7}v$, $\omega_2(t_2) = \frac{5v}{7R}$. After the time $t_1 = t_2$ the balls are moving without sliding at constant velocity and angular velocity. The graphs in question are presented in Figure 3.48.

We refer readers interested in more details to reference [30]. Part of this problem was given by the authors at the national physics competition for the fourth grade of high school in Serbia in 2012.

4

Nature

Problem 23 Allometry

Why does an elephant live longer than a mouse?
Tyrannosaurus looks like a giant chicken, so what?

Allometry is a discipline that studies the relationship of body size to shape, anatomy, and physiology. Two quantities exhibit an allometric relationship if their dependence is described using a power law. Allometry describes many phenomena in zoology using the laws of physics. We consider some of them in this problem.

(a) The mass of the skeleton of an animal m_s and the mass of the animal m are connected by an allometric relation

$$\tilde{m}_s = \alpha \tilde{m}^\beta , \tag{4.1}$$

where $\tilde{m}_s$ and $\tilde{m}$ denote dimensionless masses – that is, $\tilde{m}_s = m_s/(1\,\text{kg})$, and $\tilde{m} = m/(1\,\text{kg})$. Using the data from Table 4.1 find the constants α and β.

Table 4.1 *Dependence of skeleton mass on the mass of the animal*

Living organism	Mass [kg]	Skeleton mass [kg]
Shrewmouse	$6.30 \cdot 10^{-3}$	$3.00 \cdot 10^{-4}$
Field mouse	$2.95 \cdot 10^{-2}$	$1.30 \cdot 10^{-3}$
Cat	$8.45 \cdot 10^{-1}$	$4.36 \cdot 10^{-2}$
Rabbit	$2.20 \cdot 10^{0}$	$1.81 \cdot 10^{-1}$
Beaver	$2.27 \cdot 10^{1}$	$1.15 \cdot 10^{0}$
Human	$7.00 \cdot 10^{1}$	$1.22 \cdot 10^{1}$
Elephant	$3.50 \cdot 10^{3}$	$1.00 \cdot 10^{3}$

Paleontologists discovered the fossil remains of a tyrannosaurus and measured that the length of its step was $D = 2\,\text{m}$. The tyrannosaurus can be considered an allometrically enlarged giant chicken (the pressure in the leg bones is the same for both animals, their densities are the same, etc). The paleontologists also found part of the leg bone of a tyrannosaurus and concluded that the cross section of the bone of the tyrannosaurus was $N = 25{,}000$ times larger than the cross section of the analogous bone of the chicken.

(b) The mass of the chicken is $m_1 = 400\,\text{g}$ and the length of its leg is $l_1 = 10\,\text{cm}$. Estimate the mass of the tyrannosaurus m_2 and the length of its leg l_2. You can use the allometric relation that the ratio of the volume of the body and the third power of the length of the body is the same for all animals, as well as that the lengths of individual organs of the animal's body are proportional to the length of the animal.

(c) Estimate the natural speed of motion of the tyrannosaurus. Assume that the leg of the tyrannosaurus is a homogeneous rod and that its motion is a freely oscillating physical pendulum. Gravitational acceleration is $g = 9.81\,\text{m/s}^2$.

The approach of analyzing the functioning of living organisms using the scaling theory and allometry was started by formulating an empirical law that concerns the basal metabolic rate and its relation to the mass of the organism. The basal metabolic rate is the minimal power an organism needs to produce in order to survive. The law in question is called Kleiber's law and it is formulated as a typical allometric relation:

$$\tilde{B} = \frac{\Delta \tilde{Q}}{\Delta \tilde{t}} = \gamma \tilde{m}^{\frac{3}{4}}, \tag{4.2}$$

where $\tilde{B}$ is the dimensionless basal metabolic rate – that is, $\tilde{B} = B/(1\,\text{W})$, $\tilde{m}$ is the dimensionless mass of the animal, and γ is the dimensionless basal coefficient.

(d) Assume that the amount of energy created in the organism is proportional to the amount of food the organism consumes. Find the allometric relation between the mass of consumed food per unit mass of the animal in a day and its mass.

(e) The heart pumps blood that reaches all parts of the body. Each heartbeat introduces a certain amount of oxygen in the circulatory system. This way oxygen reaches the cells that use it during metabolic processes. For this reason the flow of oxygen in the organism is proportional to the basal metabolic rate. Find the allometric relation between the frequency of heartbeats and the mass of the animal.

(f) Assume that the heart muscle of all animals makes $n = 10^8$ contractions during the life of the animal. Find the allometric relation between the lifetime of the animal and its mass.

(g) Check how the allometric relations found in parts (d), (e), and (f) agree with real data given in Tables 4.2, 4.3, and 4.4.

Table 4.2 *Dependence of mass of consumed food per unit mass on the mass of the animal*

Living organism	Mass [kg]	$\frac{m_h}{m\Delta t}$ [1/day]
Mouse	$1.50 \cdot 10^{-2}$	280
Squirrel	$4.00 \cdot 10^{-1}$	175
Rabbit	$2.20 \cdot 10^{0}$	134
Leopard	$3.75 \cdot 10^{1}$	47
Human	$7.00 \cdot 10^{1}$	24
Zebra	$3.41 \cdot 10^{2}$	21
Elephant	$3.50 \cdot 10^{3}$	42

Table 4.3 *Dependence of frequency of heartbeat on mass of the animal*

Living organism	Mass [kg]	Heart frequency [beats/min]
Mouse	$1.50 \cdot 10^{-2}$	624
Rat	$2.26 \cdot 10^{-1}$	396
Rabbit	$2.20 \cdot 10^{0}$	210
Dog	$1.50 \cdot 10^{1}$	76
Human	$7.00 \cdot 10^{1}$	72
Tiger	$9.90 \cdot 10^{1}$	55
Donkey	$4.07 \cdot 10^{2}$	46
Elephant	$3.50 \cdot 10^{3}$	37

Table 4.4 *Dependence of the lifetime of the animal on its mass*

Living organism	Mass [kg]	Lifetime [year]
Mouse	$1.5 \cdot 10^{-2}$	3.5
Porpoise	$2.6 \cdot 10^{-1}$	7.5
Fox	$3.0 \cdot 10^{0}$	14
Goat	$3.4 \cdot 10^{1}$	18
Human	$7.0 \cdot 10^{1}$	70
Gorilla	$1.9 \cdot 10^{2}$	36
Elephant	$3.5 \cdot 10^{3}$	70

Solution of Problem 23

(a) The allometric relation $\tilde{m}_s = \alpha \tilde{m}^\beta$ can be linearized by taking the logarithm of both sides of the equation, which leads to

$$\log \tilde{m}_s = \log \alpha + \beta \log \tilde{m}. \tag{4.3}$$

Using the data from Table 4.1, in Figure. 4.1 we plot the dependence $\log \tilde{m}_s = f(\log \tilde{m})$ and the corresponding linear fit. The coefficient β is the slope of the line describing the linear fit, which leads to $\beta = 1.14$. The intersection of this line with the y-axis is $\log \alpha = -1.16$, which implies $\alpha = 0.070$.

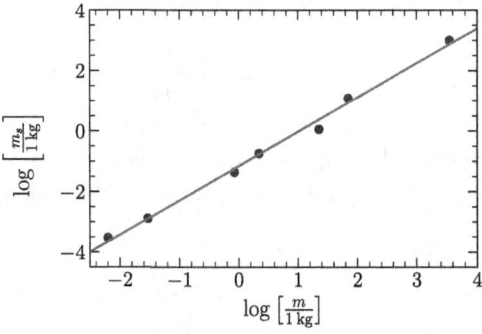

Figure 4.1 The graph of the dependence $\log \tilde{m}_s = f(\log \tilde{m})$

(b) The pressure in the leg bones of the chicken (when it is at rest standing on both legs) is the same as the pressure in the leg bones of the tyrannosaurus,

$$p = \frac{m_1 g}{2 S_1} = \frac{m_2 g}{2 S_2}, \tag{4.4}$$

where g is gravitational acceleration and S_1 and S_2 are the cross section areas of the leg bones of the chicken and the tyrannosaurus, respectively. The mass of the tyrannosaurus is

$$m_2 = m_1 \frac{S_2}{S_1} = m_1 N = 10 \,\mathrm{t}. \tag{4.5}$$

Using the allometric relation that the ratio of the volume of the body V and the third power of the length of the body L^3 is the same for all animals, as well as that the length of the leg of the animal l is proportional to the length of the body of the animal L, we obtain

$$\frac{l_2}{l_1} = \frac{L_2}{L_1} = \left(\frac{V_2}{V_1}\right)^{1/3} = \left(\frac{\rho V_2}{\rho V_1}\right)^{1/3} = \left(\frac{m_2}{m_1}\right)^{1/3} = N^{1/3}, \tag{4.6}$$

which leads to

$$l_2 = l_1 N^{1/3} = 2.92\,\text{m}. \tag{4.7}$$

(c) The natural speed of motion of the tyrannosaurus is

$$v = \frac{D}{t}, \tag{4.8}$$

where t is the time it takes for the tyrannosaurus to perform one step. If we approximate the motion of the leg with the oscillations of a physical pendulum, we have $t = T/2$, where T is the period of oscillations of the pendulum. The period of oscillations of the physical pendulum is

$$T = 2\pi \sqrt{\frac{I}{mgd}}, \tag{4.9}$$

where m is the mass of the leg, $d = l_2/2$ is the distance of the center of mass of the leg from the joint of the leg and the hip, and $I = ml_2^2/3$ is the moment of inertia. Consequently

$$t = \frac{T}{2} = \pi \sqrt{\frac{2l_2}{3g}}. \tag{4.10}$$

Using (4.8) and (4.10) we find the speed of the tyrannosaurus

$$v = \frac{D}{\pi} \sqrt{\frac{3g}{2l_2}} = 1.43\,\frac{\text{m}}{\text{s}}. \tag{4.11}$$

(d) The mass of the food a living organism consumes during the day $m_h/\Delta t$ is proportional to the energy produced in the organism during this time $\Delta Q/\Delta t$ – that is, the basal metabolic rate B, which leads to

$$\frac{m_h}{m\Delta t} \sim \frac{B}{m}. \tag{4.12}$$

The dimensionless form of the relation in equation (4.12) together with equation (4.2) yields

$$\frac{m_h}{m\Delta \tilde{t}} \sim \tilde{m}^{-1/4}. \tag{4.13}$$

(e) The mass of oxygen m_{O_2} that is brought into the circulatory system during the time interval T_s of one heartbeat is proportional to the basal metabolic rate B - that is,

$$\frac{m_{O_2}}{T_s} \sim B. \tag{4.14}$$

The mass of oxygen m_{O_2} is proportional to the mass of the heart m_{sr}, which is further proportional to the mass m, which implies

$$f = \frac{1}{T_s} \sim \frac{B}{m}. \tag{4.15}$$

Dimensionless form of the relation in equation (4.15) together with equation (4.2) yields

$$\tilde{f} \sim \tilde{m}^{-1/4}. \tag{4.16}$$

(f) If the heart muscle of all animals performs n contractions during their life, the lifetime of the animal t is given as $t = nT_s$, which, using equations (4.15) and (4.16), leads to

$$\tilde{t} \sim \tilde{m}^{1/4}. \tag{4.17}$$

(g) A linear fit (solid line) to the data from Table 4.2 (circles) is presented in Figure 4.2. The slope of the linear fit -0.21 is very close to the value -0.25 expected from equation (4.13).

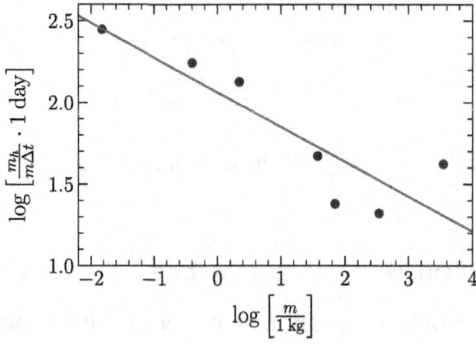

Figure 4.2 The graph of the dependence $\log\left(\frac{m_h}{m\Delta t}\right) = f(\log \tilde{m})$

A linear fit (solid line) to the data from Table 4.3 (circles) is presented in Figure 4.3. The slope of the linear fit is -0.25, which is the same value obtained in part (e).
A linear fit (solid line) to the data from Table 4.4 (circles) is presented in Figure 4.4. The slope of the linear fit is 0.25, which is the value expected based on equation (4.17).

We refer interested readers to references [20] and [40]. The authors presented part of this problem at the national physics competition for the third grade of high school in Serbia in 2016.

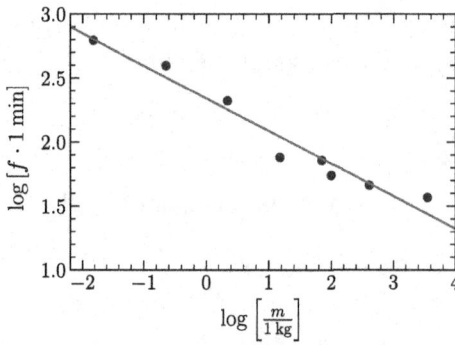

Figure 4.3 The graph of the dependence $\log \tilde{f} = f(\log \tilde{m})$

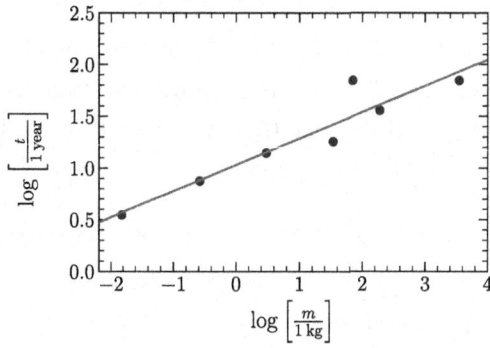

Figure 4.4 The graph of the dependence $\log \tilde{t} = f(\log \tilde{m})$

Problem 24   Walking on the Surface of Water

Why can insects stand on the surface of water while humans cannot?
How do some insects move on the surface of water?

Unlike humans, many insects are capable of walking on the surface of water without even getting wet. We investigate in this problem why this is the case.

In all parts of the problem you may assume that the density of water is $\rho_v = 1,000\ \frac{\text{kg}}{\text{m}^3}$ and gravitational acceleration is $g = 9.81\ \frac{\text{m}}{\text{s}^2}$.

(a) Humans are capable of floating on the surface of water, but in this case the largest part of the body is immersed in water. At the same time humans need to perform characteristic moves with their arms and legs in order to remain in balance. Consider a woman of mass $m = 50.0\,\text{kg}$ in a vertical position in which $f = 95.0\%$ of her volume is immersed in water, as shown in Figure 4.5. To remain in this position, the woman moves her legs in such a way that particles of water gain velocity directed downward. Assume that her leg movements are such that in a small time interval Δt all particles of water in a cylinder of height $v\Delta t$ and area $S_{\text{eff}} = 600\,\text{cm}^2$ (shown in Figure 4.5) gain the speed v. The density of the woman ρ_z is equal to the density of water ρ_v. Determine the speed v, as well as the power the woman imparts to the particles of water.

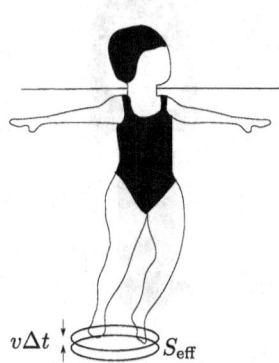

Figure 4.5 The woman in balance on the surface of water

(b) For which value of v will the woman stand on the water in the sense that a negligible part of her body will be immersed in water? What power does the woman impart to the water particles in this case?

(c) The woman considered in previous parts of this problem is a water polo goalkeeper. Many water polo goalkeepers have the ability to keep a large part of their body above the water surface (Figure 4.6). This way they have a better chance of defending their goal from the opponent's shots. What power should this woman impart to the particles of water so that only $f = 50.0\%$ of her body remains immersed in water?

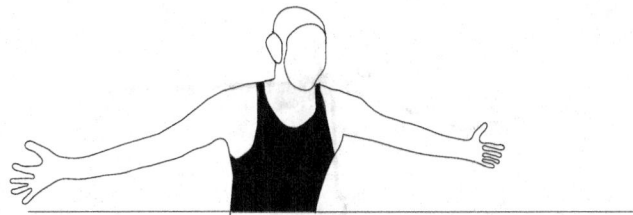

Figure 4.6 A woman as a water polo goalkeeper in the moment when a large part of her body is above the surface of water

In parts (a)–(c) we considered one way a woman can remain in balance on the surface of water. On the other hand, capillary effects can also act to keep a certain object on the surface of water.

(d) Determine the ratio of the maximal vertical component of the capillary force that can act on the woman when she stands on the surface of water (as shown in Figure 4.7) and the gravitational force. The circumference of the woman's foot is $\mathcal{O} = 50.0\,\text{cm}$, while the surface tension of water is $\sigma = 0.0700\,\frac{\text{N}}{\text{m}}$.

Figure 4.7 A woman standing on the surface of water

Unlike humans, certain insects are capable of walking on the surface of water. These insects typically have long and thin legs whose bottom half is in contact with the surface of water; see Figure 4.8.

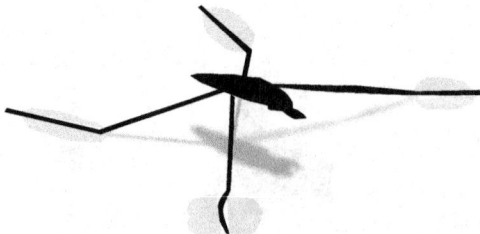

Figure 4.8 An insect on the surface of water. Black surfaces denote the insect, while gray surfaces denote the shadow of the insect on the surface of water

(e) We consider first how a part of one insect leg interacts with the surface of water. We approximate this part of the leg with a stick in the shape of a cylinder with base radius r and a length much longer than r. The density of the stick ρ_{in} is equal to the density of water. Due to interaction of the stick with the surface of water, deformation of the surface of water occurs, as shown in Figure 4.9. The point in which the stick loses contact with the surface of water is determined by the angle $\alpha = 45°$, which is shown in Figure 4.9. Determine the value of r.

Figure 4.9 Cross section of the part of an insects's leg that is contact with water and the shape of the water surface

(f) Determine the maximal size of the insect l that still allows the insect to stand on the surface of water due to surface tension, as shown in Figure 4.8. Assume that the insect has four legs, that the length of the part of the leg that is in contact with the water is l, that the volume of the insect is $V = l^3$, and that the density of the insect is $\rho_{in} = \rho_v$.

Some insects also use capillary effects to accelerate their motion. These insects release a substance that modifies the surface of water behind them. Consequently the surface tension of the fluid in contact with the rear part of the leg is $\sigma' = 0.0500 \frac{N}{m} < \sigma$ (see Figure 4.10).

(g) A four-legged insect starts from rest and moves on the surface of water using this effect (called the Marangoni effect). The direction of motion is shown in Figure 4.10. The angle between this direction and the direction of the parts of the front legs that are in contact with water is 45°. What is the acceleration of this insect, and what will be its speed $t = 0.500$ s after the beginning of motion? The mass of the insect is $m = 0.0300$ g, while the length of the part of the leg that is in contact with the water is $l = 0.750$ mm. Neglect all resistance forces.

Solution of Problem 24

(a) In small time Δt the woman acts on the particles of water inside a volume $\Delta V = S_{eff} v \Delta t$. The mass of these particles is

$$\Delta m = \rho_v \Delta V = \rho_v S_{eff} v \Delta t. \tag{4.18}$$

The woman acts on the particles of water by the force $F = \frac{\Delta p}{\Delta t}$, where $\Delta p = \Delta m v$ is the momentum imparted to the particles of water. Previous equations lead to

$$F = \rho_v S_{eff} v^2. \tag{4.19}$$

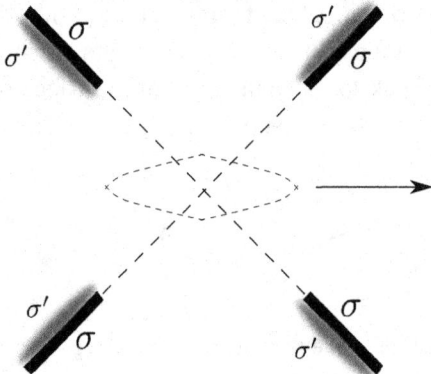

Figure 4.10 The projection of an insect on the surface of water. The black color denotes the parts of the leg in contact with the water. The areas where the surface tension is σ and the areas where it is equal to σ' are also denoted. The arrow denotes the direction of the motion of the insect.

Newton's third law implies that the particles of water act on the woman by the force of the same magnitude that is directed upward. The force of gravity whose magnitude is mg and the buoyancy force whose magnitude is

$$F_p = \rho_v fV g \tag{4.20}$$

also act on the woman, where

$$fV = f\frac{m}{\rho_z} \tag{4.21}$$

is the volume of the part of the woman's body immersed in water. When the woman is in equilibrium, it follows that

$$mg = F_p + F. \tag{4.22}$$

From equations (4.19)–(4.22) it follows that

$$v = \sqrt{\frac{m\left(1 - f\frac{\rho_v}{\rho_z}\right)g}{\rho_v S_{\text{eff}}}} = 0.639\,\frac{m}{s}. \tag{4.23}$$

The power the woman imparts to the particles of water is equal to the change of their kinetic energy in unit time

$$P = \frac{\Delta E_k}{\Delta t}. \tag{4.24}$$

Since

$$\Delta E_k = \frac{1}{2}\Delta m v^2, \tag{4.25}$$

from equations (4.18), (4.24), and (4.25) it follows that

$$P = \frac{1}{2}\rho_v S_{\text{eff}} v^3 = 7.84 \, \text{W}. \tag{4.26}$$

(b) The speed in question is given by equation (4.23), where $f = 0$. Therefore, $v = 2.86 \, \frac{\text{m}}{\text{s}}$. The power the woman imparts to the particles of water is given by equation (4.26), which leads to $P = 701 \, \text{W}$. This power is larger than the power a human can produce. Therefore, humans cannot stand on the surface of water.

(c) The speed v is given by equation (4.23), where $f = 0.5$, which leads to $v = 2.02 \, \frac{\text{m}}{\text{s}}$. The power the woman imparts to the particles of water is given by equation (4.26), which leads to $P = 248 \, \text{W}$. This power is comparable to the power a human can produce. Therefore, water polo goalkeepers can keep a significant part of their body above the surface of water.

(d) The vertical component of the capillary force is $F_v = 2\sigma \, \mathcal{O} \cos \theta$, where θ is the angle between the direction of this force and the vertical. The maximal value of this component is reached for $\theta = 0$ and is equal to $F_v^{\text{max}} = 2\sigma \mathcal{O}$. The ratio in question is $x = \frac{F_v^{\text{max}}}{mg} = \frac{2\sigma \mathcal{O}}{mg} = 1.43 \cdot 10^{-4}$. From this result we see that the capillary force is much smaller than the gravity force. Therefore, the woman cannot stand on the surface of water.

(e) The forces acting on the stick are shown in Figure 4.11. The vertical component of the capillary force is

$$F_v = 2\sigma l \sin \alpha, \tag{4.27}$$

where l is the length of the stick. The buoyancy force acting on the stick is

$$F_p = \rho_v V' g, \tag{4.28}$$

where V' is the volume of the stick below the horizontal plane determined by the points where the stick loses contact with the water. Since the density of the stick is equal to the density of water, the gravity force is

$$mg = \rho_{\text{in}} V g = \rho_v V g, \tag{4.29}$$

where V is the volume of the stick. Using $\frac{V}{l} = r^2 \pi$ and $\frac{V'}{l} = r^2 \left(\alpha - \frac{1}{2} \sin 2\alpha \right)$, from equations (4.27), (4.28), and (4.29) and the condition $mg = F_v + F_p$ we obtain

$$r = \sqrt{\frac{2\sigma \sin \alpha}{\rho_v g \left(\pi - \alpha + \frac{1}{2} \sin 2\alpha \right)}} = 1.88 \, \text{mm}. \tag{4.30}$$

(f) The maximal vertical component of the buoyancy force acting on one leg is $F_v = 2\sigma l$. Since the insect has four legs, the total vertical component is $F_{\text{tot}} = 4F_v = 8\sigma l$. The insect can stand on the surface of water when this force is larger

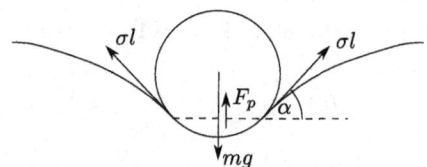

Figure 4.11 With the solution of problem 24(e)

than the gravity force $mg = \rho_{in} l^3 g = \rho_v l^3 g$. From the condition $F_{tot} > mg$ it follows that $l < \sqrt{\dfrac{8\sigma}{\rho_v g}} = 7.56\,\text{mm}$.

(g) The vertical component of the capillary force on one leg is $F_v = \sigma l \sin\theta + \sigma' l \sin\theta$ (see Figure 4.12), while its projection to the horizontal plane is $F_h = \sigma l \cos\theta - \sigma' l \cos\theta$. The projection to the plane of the insect's motion is $F_k = F_h \dfrac{\sqrt{2}}{2}$. Since the insect does not move in a vertical direction, it follows that $mg = 4F_v$. Acceleration of the insect is obtained from $ma = 4F_k$. Using previous equations and the identity $\cos\theta = \sqrt{1 - \sin^2\theta}$, which is valid for angles $\theta \in [0, \pi/2]$, we obtain

$$a = \frac{2\sqrt{2}(\sigma - \sigma')l}{m}\sqrt{1 - \frac{m^2 g^2}{16 l^2 (\sigma + \sigma')^2}} = 0.814\,\frac{\text{m}}{\text{s}^2}. \qquad (4.31)$$

This leads to $v = at = 0.407\,\frac{\text{m}}{\text{s}}$.

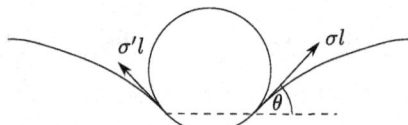

Figure 4.12 With the solution of problem 24(g): the directions of the capillary forces acting on one leg of the insect

We refer interested readers to references [5] and [9]. The authors gave part of this problem at a regional physics competition for the fourth grade of high school in Serbia in 2017.

Problem 25   Colors of Doves and Butterflies

Why do we see green and purple colors on a dove's neck?

Why does the color of some butterfly wings change when viewed at a different angle?

The colors of many butterflies and birds are caused by the effects of interference and diffraction of light on the structures present on the surface of their body. As a

consequence of these effects, reflected light of a certain wavelength is more inten-
sive and the observer sees the color that corresponds to this wavelength. Assume
in this problem that the range of wavelengths of visible light is 400 to 700 nm,
where the range 400 to 450 nm corresponds to purple, 450 to 510 nm to blue, 510
to 560 nm to green, 560 to 620 nm to yellow, and 620 to 700 nm to red.

You have probably noticed that the neck of a dove contains areas of green
and purple color. In this area the dove's feather contains thin barbules that can
be modeled as a plate of thickness $d = 650$ nm and refractive index $n = 1.50$
shown in Figure 4.13(b). Consider a plane monochromatic wave whose direction
of propagation is perpendicular to the surface of the barbule.

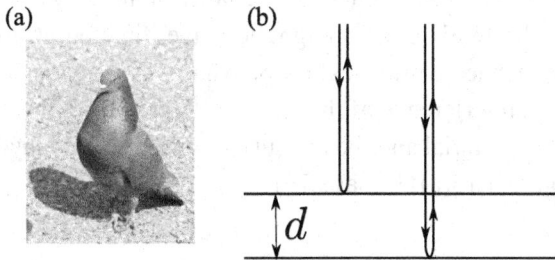

Figure 4.13 (a) The dove. Reprinted with permission from reference [19]. (b)
Scheme of the ray reflected from top surface of the barbule and the ray reflected
from bottom surface of the barbule

(a) For what wavelengths will the condition for constructive interference of the
rays reflected from top and bottom surface of the barbule (shown in Figure 4.13[b]) be satisfied?
(b) Calculate the wavelengths from the visible part of the spectrum that satisfy the
condition derived in part (a).
(c) Using the solution of previous parts of the problem explain why we see green
and purple colors on the dove's neck.

One of the most investigated butterflies is the morpho butterfly, shown in Figure 4.14(a). The scales on the wings of this butterfly contain complex structures
shown in Figure 4.14(b) that resemble bookshelves from a library. A consequence of such structures is their blue color when the direction of their observation
is perpendicular to their surface (Figure 4.14[c]), and their purple color when
the direction of observation is at an angle of $45°$ with respect to their surface
(Figure 4.14[d]).

(d) In this part of the problem we approximate the complex structure of the scales
by a periodic array of layers of the material with refractive index of $n_1 = 1.53$

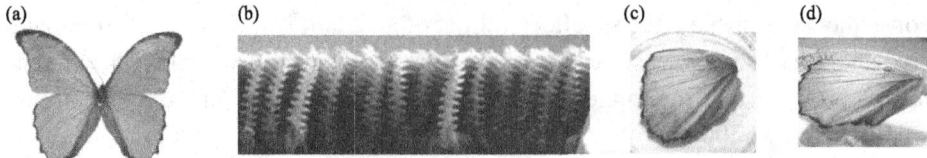

Figure 4.14 (a) Morpho butterfly. (b) The structure of the scales of the morpho butterfly. (c) Image of the wings observed at a right angle with respect to their surface. (d) Image of the wings observed at an angle of $45°$ with respect to their surface. Reprinted with permission from reference [19]

and the air, shown in Figure 4.15(a). The thickness of the layer of the material is $d_1 = 65.0$ nm, while the thickness of the layer of air is $d_2 = 130$ nm. Consider a plane monochromatic electromagnetic wave directed at a right angle with respect to the surface of the scale. For what wavelengths will the condition for constructive interference of the waves reflected from the top surface of the first layer of the material and the top surface of the second layer of the material [shown in Figure 4.15(a)] be satisfied?

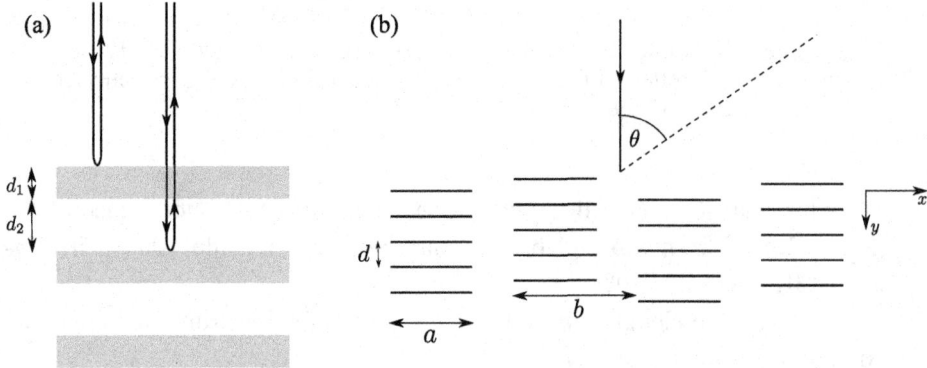

Figure 4.15 (a) The model of the scale as a periodic array of layers of the material with refractive index n_1 and the air. (b) A more complex model of the scale

(e) Using the solution of part (d) explain why we see a blue color on the butterfly when we observe it at a right angle with respect to the surface of the wings.

Although the model considered in part (d) can explain why the butterfly is blue when we observe it at a right angle, the model cannot fully explain the change in color when the angle of observation is changed. Therefore we consider a more complex model of the scale shown in Figure 4.15(b). In this model, the scale consists of an array of thin parallel plates of width a. The distance between neighboring plates inside one column is d. The length of the plate in the direction perpendicular to the plane of the figure is much longer than a, d, and the wavelength of light. Two neighboring columns are shifted in the x-direction by b, while the i-th column is shifted in the y-direction by y_i. y_i is a random variable whose mean value is zero and whose variations are much larger than the wavelength of light. The number of plates in one column is M, while the number of columns is N.

(f) Consider first the interaction of light with one plate of width a. The plane mono-chromatic electromagnetic wave of wavelength λ_0 was directed at a right angle with respect to its surface; see Figure 4.16(a). A remote observer whose position is such that the angle between the direction of incident and the reflected ray is θ measures the intensity of the wave $I_1(\theta) = c_1 f_1(\theta)$, where c_1 is a constant and $f_1(\theta)$ is a function that may contain the constants a and λ_0. Determine the function $f_1(\theta)$.

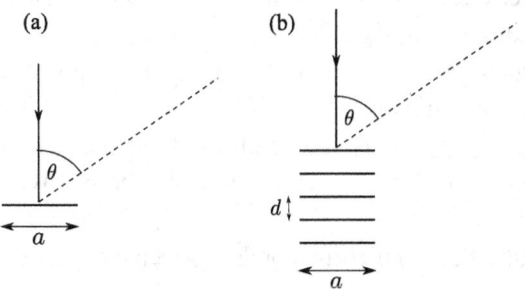

Figure 4.16 (a) One plate on the scale. (b) A column of plates on the scale

(g) We consider the interaction of light with one column of the plates; see Figure 4.16(b). Assume that the intensity of light changes very little after passage through one plate – that is, that the transmission coefficient through the plate is close to 1 and much larger than the coefficients of reflection and absorption. The light was directed to the column of plates in the same manner as in part (f) and the intensity of the reflected light $I_2(\theta)$ is measured in the same manner. Show that $I_2(\theta) = c_2 f_1(\theta) f_2(\theta)$, where c_2 is a constant, $f_1(\theta)$ is a function determined in part (f), and $f_2(\theta)$ is a function that should be determined and that may contain the constants d, M, and λ_0.

(h) We finally consider the interaction of light with the whole structure of the scale, shown in Figure 4.15(b). Show that the intensity of reflected light is given by the expression $I_3(\theta) = c_3 f_1(\theta) f_2(\theta) f_3(\theta)$, where $f_1(\theta)$ and $f_2(\theta)$ are the functions determined in previous parts of the problem, $f_3(\theta)$ is a function that should be determined and that may contain the constants b, N, and λ_0, and c_3 is a constant. The light was directed toward the structure in the same manner as in part (f), and the intensity of reflected light is measured in the same manner.

(i) The parameters of the structure are $a = 300\,\text{nm}$, $d = 235\,\text{nm}$, and $M = 9$. Plot the graph of the function $f(\theta) = f_1(\theta) f_2(\theta) f_3(\theta)$ for $\lambda_0 = 480\,\text{nm}$ and $\lambda_0 = 410\,\text{nm}$.

(j) Using the solution of part (i) explain why we see blue when we observe the wing at a right angle with respect to its surface and why we see purple when we observe it at an angle around $45°$.

Solution of Problem 25

(a) The phase of the ray changes by $\phi_1 = \pi$ when it reflects on the top surface of the scale because it reflects from the optically denser medium. Double passage of the ray through the scale and reflection from the bottom surface change its phase by $\phi_2 = \frac{2\pi}{\lambda}2nd$, where λ is the wavelength of the wave in air. Constructive interference occurs when $\phi_2 - \phi_1 = 2m\pi$, where m is a nonnegative integer. We finally obtain $\lambda = \frac{4nd}{2m+1}$.
Note: The structure considered in this part of the problem is known as the Fabry–Perot interferometer.

(b) The wavelengths that satisfy the condition from part (a) are from the visible part of the spectrum when $m = 3$ and $m = 4$. These wavelengths are 557 nm and 433 nm.

(c) The wavelength of 557 nm corresponds to green, while 433 nm corresponds to purple. For this reason we see green and purple areas on the dove's neck.
Note: This explanation of the origin of color on the dove's neck is somewhat simplified. Detailed research of the origin of these colors is presented in reference [39].

(d) The phase of the ray changes by $\phi_1 = \pi$ when it reflects on the top surface. Double passage of the ray through the first layer of the material and the first layer of air and reflection on the second layer of material lead to a phase change of $\phi_2 = \frac{2\pi}{\lambda}2(n_1 d_1 + d_2) + \pi$. Constructive interference occurs when $\phi_2 - \phi_1 = 2m\pi$, where m is a positive integer. This leads to $\lambda = \frac{2(n_1 d_1 + d_2)}{m}$.
Note: The structure considered in this part of the problem is known as the distributed Bragg reflector.

(e) The wavelength that satisfies the condition from part (d) is from the visible spectrum when $m = 1$ and it is equal to $\lambda = 459$ nm. For this reason, we see blue when we observe the butterfly at a right angle with respect to its surface.

(f) We choose the coordinate system as in Figure 4.17. The n-axis is in the direction of the reflected wave that reaches the observer. The contribution to the relevant component of the electric field due to reflection on the part of the plate with coordinates from interval $(x, x+dx)$ is $dE = C \cdot dx \cdot \cos[\omega t - kn + \phi(x)]$, where C is a constant that depends on the intensity of the incident wave and the reflection coefficient on the plate, ω is the angular frequency of the wave, t is the time, $k = \frac{2\pi}{\lambda_0}$ is the wave vector, and $\phi(x)$ is the phase of the wave. Since the phase of the wave reflected at x_0, where x_0 is in the middle of the plate, is delayed with respect to the phase of the wave reflected at x by $\frac{2\pi}{\lambda_0}\overline{AC}$ (see Figure 4.17), we have $\phi(x) = \phi(x_0) + \frac{2\pi}{\lambda_0}(x - x_0)\sin\theta$. The component of the electric field of the reflected wave is then $E(n,t) = \int_{x_0 - \frac{a}{2}}^{x_0 + \frac{a}{2}} C \cdot dx \cdot \cos[\omega t - kn + \phi(x)]$. By solving the integral, we obtain

$$E(n,t) = Ca\frac{\sin\left(\frac{\pi a}{\lambda_0}\sin\theta\right)}{\frac{\pi a}{\lambda_0}\sin\theta}\cos[\omega t - kn + \phi(x_0)].$$

From the previous expression and the fact that the intensity of light is proportional to the square of the amplitude we find

$$f_1(\theta) = \frac{\sin^2\left(\frac{\pi a}{\lambda_0}\sin\theta\right)}{\left(\frac{\pi a}{\lambda_0}\sin\theta\right)^2}.$$

Note: The effect considered in this part of the problem is known as the single-slit diffraction.

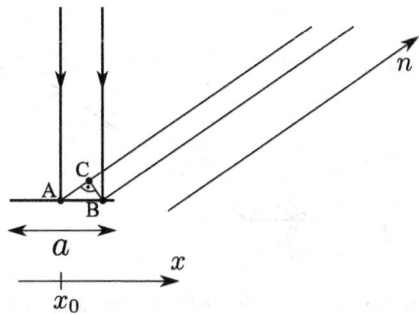

Figure 4.17 Reflection of light on one plate

(g) The component of the electric field of the reflected wave at the position of the remote observer is obtained by adding the contributions of waves reflected at each of the plates. The phase difference of the waves reflected from two neighboring plates is $\Delta\phi = \frac{2\pi}{\lambda_0}(\overline{AC} + \overline{AB})$ (see Figure 4.18) – that is, $\Delta\phi = \frac{2\pi}{\lambda_0}(d + d\cos\theta)$. The component of the electric field is then

$$E_t(n,t) = Ca\frac{\sin\left(\frac{\pi a}{\lambda_0}\sin\theta\right)}{\frac{\pi a}{\lambda_0}\sin\theta}\sum_{i=0}^{M-1}\cos[\omega t - kn + \phi(x_0) - i\Delta\phi].$$

Using the identity

$$\sum_{j=0}^{K} \cos(\alpha + j\beta) = \frac{\sin \frac{(K+1)\beta}{2}}{\sin \frac{\beta}{2}} \cos\left(\alpha + \frac{K\beta}{2}\right),$$

we obtain

$$E_t(n,t) = Ca \frac{\sin\left(\frac{\pi a}{\lambda_0}\sin\theta\right)}{\frac{\pi a}{\lambda_0}\sin\theta} \frac{\sin\left[\frac{M\pi d}{\lambda_0}(1+\cos\theta)\right]}{\sin\left[\frac{\pi d}{\lambda_0}(1+\cos\theta)\right]} \cos\left[\omega t - kn + \phi(x_0) - \frac{M-1}{2}\Delta\phi\right].$$

We find from the previous expression that

$$I_2(\theta) \propto \frac{\sin^2\left(\frac{\pi a}{\lambda_0}\sin\theta\right) \sin^2\left[\frac{M\pi d}{\lambda_0}(1+\cos\theta)\right]}{\left(\frac{\pi a}{\lambda_0}\sin\theta\right)^2 \sin^2\left[\frac{\pi d}{\lambda_0}(1+\cos\theta)\right]},$$

which leads to

$$f_2(\theta) = \frac{\sin^2\left[\frac{M\pi d}{\lambda_0}(1+\cos\theta)\right]}{\sin^2\left[\frac{\pi d}{\lambda_0}(1+\cos\theta)\right]}.$$

Note: The structure considered in this problem is one type of diffraction grating.

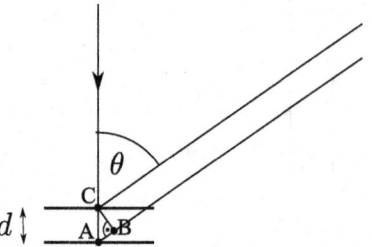

Figure 4.18 Reflection of light from two neighboring plates

(h) The component of the electric field of the reflected wave is obtained by adding the contributions of each of the column of plates. Since the displacement of the i-th column of the plate in the y-direction is a random variable, the phase of the reflected wave will also be a random variable. For this reason, the intensity of the reflected wave is equal to the sum of the intensities of the waves reflected from each of the column of plates – that is, $I_3(\theta) = NI_2(\theta)$. Since the constant N can be included in the definition of the proportionality constant c_3, it follows that $f_3(\theta) = 1$.

(i) We present in Figure 4.19 the dependence $f(\theta)$ for $\lambda_0 = 410$ nm and $\lambda_0 = 480$ nm.

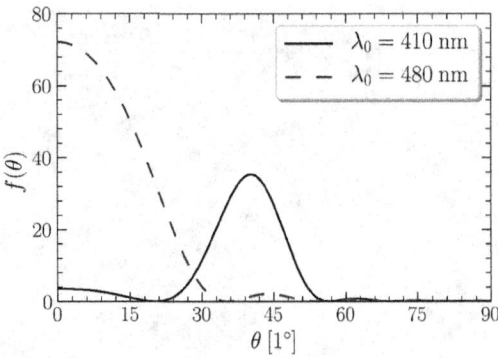

Figure 4.19 Angular dependence of the intensity of reflected light $I(\theta) \propto f(\theta)$ for $\lambda_0 = 410$ nm and $\lambda_0 = 480$ nm

(j) From the solution of part (i), we find that the maximum of reflected light for $\lambda_0 = 480$ nm is at an angle of $\theta = 0°$, while the maximum is at $\theta = 40°$ for $\lambda_0 = 410$ nm. For this reason, we see blue when we observe the scale at an angle of $\theta = 0°$, while we see purple when we observe it at an angle of around $45°$.

We refer the interested reader to references [19] and [39]. The authors presented a modified version of this problem at the Serbian Physics Olympiad for high school students in 2019.

Problem 26 Rainbow

Why do we see a rainbow?
At what angles do we see a rainbow?
Why does the sun have to be behind the observer when the observer sees the rainbow?

A rainbow is a natural phenomenon caused by refraction of sunrays at raindrops and their reflection within the raindrops. If the sun is behind the observer, the observer sees in the sky circular arcs of all visible colors. The rainbow can also be seen by observing a fountain on a sunny day, as shown in Figure 4.20.

In this problem we investigate in detail how a rainbow appears. Assume that all raindrops have an ideally spherical shape. The radius of the sphere is much larger than the wavelength of light. The refractive index of water for light whose wavelength in the air is $\lambda_1 = 400$ nm (purple), $\lambda_2 = 550$ nm (green) and $\lambda_3 = 700$ nm (red) is, respectively, $n_1 = 1.339$, $n_2 = 1.333$ and $n_3 = 1.331$.

(a) A parallel beam of light is incident on a raindrop. Consider a ray that exhibits one internal reflection at the surface of the raindrop and enters the raindrop in

Figure 4.20 The appearance of a rainbow in a fountain at the lake in Canberra, Australia

point A, where $\angle AOB = \alpha$ (Figure 4.21) and the direction OB is parallel to the direction of incident rays. Let θ_p be the angle between the horizontal and the ray that leaves the raindrop. Find the expression for θ_p in terms of α and refractive index n.

(b) Plot the graph of the dependence of θ_p on α for green light. Calculate α for which θ_p is maximal and calculate the maximal value of θ_p.

(c) The sun is behind the observer, as shown in Figure 4.22. Use previous results to calculate the angle θ between the direction of the observer's view and the horizontal when the observer sees the green arc of the primary rainbow that forms from the rays that exhibited one internal reflection in the raindrop.

(d) A secondary rainbow forms from rays that exhibited two internal reflections in the raindrop, as shown in Figure 4.23. Let θ_s be the angle between one such ray and the horizontal when the ray leaves the raindrop. Express θ_s as a function

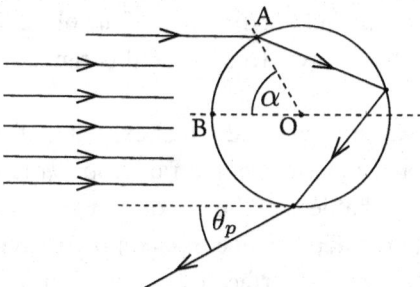

Figure 4.21 Refraction and internal reflection of a ray on the raindrop. Point O is the center of the sphere.

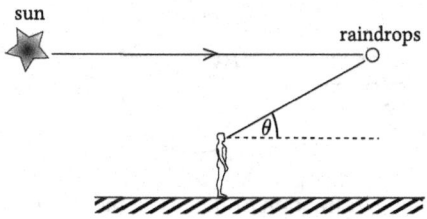

Figure 4.22 The position of the sun, the observer, and the raindrops

of α and n. Plot the graph of the dependence of θ_s on α for green light and calculate the angle θ between the direction of the view of the observer from part (c) and the horizontal when the observer sees the green arc of the secondary rainbow.

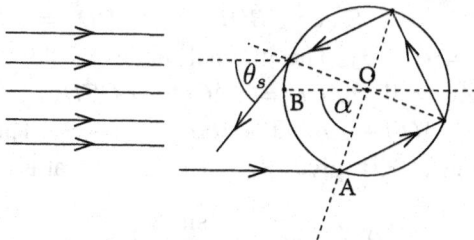

Figure 4.23 Refraction and internal reflections of the ray that forms the secondary rainbow

(e) Determine the angular widths of the primary and secondary rainbows. Angular width is defined as the width of the range of angles for which the observer from part (c) sees some of the colors from the visible spectrum $\lambda \in (400, 700)$ nm.

(f) Assume that only the rays that exhibited one or two internal reflections in the raindrop reach the observer from part (c). For a certain range of angles

θ between the horizontal and the direction of the observer's view, the observer does not see any of these rays. Determine this range of angles (the so-called Alexander's band).

(g) Consider a ray, shown in Figure 4.24, that exhibits refraction at the surface of the raindrop twice and then leaves the raindrop. θ_0 is the angle between the horizontal and the ray that leaves the raindrop. Express θ_0 in terms of α and n and plot the graph of the dependence of θ_0 on α for green light. Can such rays (that exhibit refraction at the surface of the raindrop twice and then leave the drop) form the rainbow for some position of the observer and the sun?

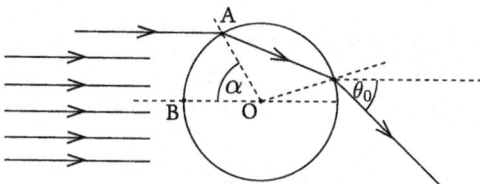

Figure 4.24 Double refraction of the ray at the surface of the raindrop

Solution of Problem 26

(a) The angle between a ray that refracts at point A and a line OA that is perpendicular to the surface of the raindrop is α. Snell's law yields $\sin \alpha = n \sin \beta$, where $\beta = \angle OAC$ (Figure 4.25). The triangle AOC is isosceles and therefore $\angle OCA = \angle OAC = \beta$. The law of reflection at point C gives $\angle DCO = \angle OCA = \beta$. Since the triangle DOC is also isosceles, we have $\angle ODC = \angle DCO = \beta$. Snell's law at point D yields $n \sin \beta = \sin \angle D_3 D D_4$ and consequently $\angle D_3 D D_4 = \alpha$. Since $\angle BOD = \angle D_2 D D_4 = \angle D_2 D D_3 + \angle D_3 D D_4$, we have $\angle BOD = \alpha + \theta_p$. The internal angles of the quadrilateral AODC are equal to $\angle OAC = \beta$, $\angle ACD = \angle ACO + \angle OCD = 2\beta$, $\angle ODC = \beta$, and $\angle AOD = 360° - \angle AOB - \angle BOD = 360° - 2\alpha - \theta_p$. The condition that the sum of these angles is $360°$ gives $\theta_p = 4\beta - 2\alpha$ – that is,

$$\theta_p = 4 \arcsin\left(\frac{\sin \alpha}{n}\right) - 2\alpha. \tag{4.32}$$

(b) The graph of the dependence of θ_p on α calculated using equation (4.32) for green light ($n = 1.333$) is given in Figure 4.26. The angle α for which this dependence exhibits a maximum can be read from the graph or it can be determined from the condition $d\theta_p/d\alpha = 0$. Since

$$\frac{d\theta_p}{d\alpha} = \frac{\frac{4\cos \alpha}{n}}{\sqrt{1 - \left(\frac{\sin \alpha}{n}\right)^2}} - 2, \tag{4.33}$$

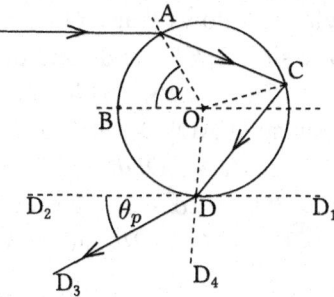

Figure 4.25 With the solution of problem 26(a)

the angle α for which θ_p is maximal reads

$$\alpha = \arcsin \sqrt{\frac{4 - n^2}{3}} = 59.4°. \tag{4.34}$$

It follows that maximal θ_p is $\theta_p^{\max} = 42.1°$.

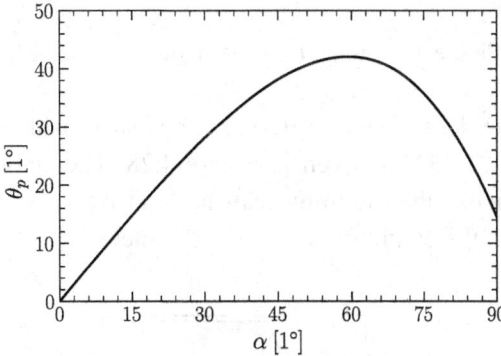

Figure 4.26 With the solution of problem 26(b)

(c) We see from the graph presented in Figure 4.26 that the largest number of rays reaches the observer under the angles around $\theta = \theta_p^{\max} = 42.1°$. Therefore, the direction of the view of the observer that sees the rainbow is exactly determined by this angle.

(d) The angle between a ray that refracts at point A and a line OA that is perpendicular to the surface of the raindrop is α. Snell's law yields $\sin\alpha = n\sin\beta$, where $\beta = \angle OAC$ (Figure 4.27). The triangle AOC is isosceles and therefore $\angle OCA = \angle OAC = \beta$. The law of reflection at point C gives $\angle DCO = \angle OCA = \beta$. Since the triangle DOC is isosceles, we have $\angle ODC = \angle DCO = \beta$. The law of reflection at point D yields $\angle CDO = \angle ODE = \beta$ and from the fact that the triangle DOE is isosceles it follows that $\angle ODE = \angle DEO = \beta$. Snell's law at point E gives $n\sin\beta = \sin\angle E_1EE_2$, and consequently $\angle E_1EE_2 = \alpha$. Since $\angle BOE = \angle E_3EE_1 = \angle E_2EE_1 - \angle E_2EE_3$, it follows $\angle BOE = \alpha - \theta_s$. The internal angles in the pentagon AOEDC are $\angle OAC = \beta$, $\angle ACD = \angle ACO + \angle OCD = 2\beta$, $\angle CDE = \angle CDO + \angle ODE = 2\beta$, $\angle OED = \beta$, and $\angle EOA = 360° - \angle BOE - \angle AOB = 360° - 2\alpha + \theta_s$. The condition that the sum of these angles is $540°$ gives $\theta_s = 180° - 6\beta + 2\alpha$ and consequently

$$\theta_s = 180° - 6\arcsin\left(\frac{\sin\alpha}{n}\right) + 2\alpha. \tag{4.35}$$

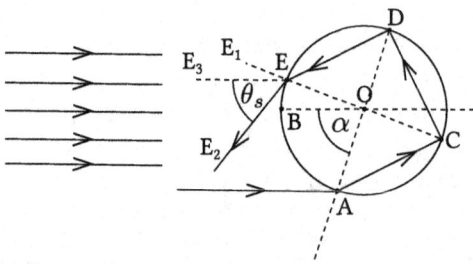

Figure 4.27 With the solution of problem 26(d)

The graph of the dependence of θ_s on α calculated using equation (4.35) for green light ($n = 1.333$) is given in Figure 4.28. The angle α for which this dependence reaches the minimum can be read from the graph or it can be determined from the condition $d\theta_s/d\alpha = 0$. Since

$$\frac{d\theta_s}{d\alpha} = \frac{-\frac{6\cos\alpha}{n}}{\sqrt{1 - \left(\frac{\sin\alpha}{n}\right)^2}} + 2, \tag{4.36}$$

it follows that the angle α for which θ_s reaches minimum reads

$$\alpha = \arcsin\sqrt{\frac{9 - n^2}{8}} = 71.8°. \tag{4.37}$$

It then follows that the minimal θ_s is $\theta_s^{\min} = 50.9°$. The largest number of rays that exhibit two internal reflections reach the observer under the angles around $\theta = \theta_s^{\min} = 50.9°$. Therefore, the direction of the view of the observer who sees the secondary rainbow is determined by this angle.

(e) From equations (4.34) and (4.32) we find that in the case of red light $\theta_p^{\max}(n_3) = 42.4°$, and in the case of purple light $\theta_p^{\max}(n_1) = 41.2°$. The angular width of the primary rainbow is therefore $\Delta\theta_p = \theta_p^{\max}(n_3) - \theta_p^{\max}(n_1) = 1.2°$. Using equations (4.37) and (4.35), we find that $\theta_s^{\min}(n_3) = 50.4°$ for red light and

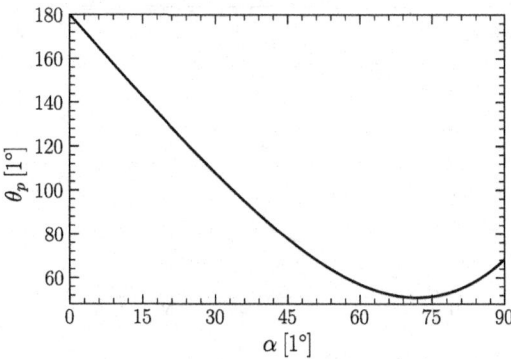

Figure 4.28 With the solution of problem 26(d)

$\theta_s^{\min}(n_1) = 52.5°$ for purple light. The angular width of the secondary rainbow is therefore $\Delta\theta_s = \theta_s^{\min}(n_1) - \theta_s^{\min}(n_3) = 2.1°$.

(f) Using the solutions of previous parts of the problem, we conclude that in the range of angles from $42.4°$ to $50.4°$ the observer does not register any of the rays in question.

(g) The angle between a ray that refracts at point A and a line OA that is perpendicular to the surface of the raindrop is α. Snell's law yields $\sin\alpha = n\sin\beta$, where $\beta = \angle OAC$ (Figure 4.29). The triangle AOC is isosceles, which gives $\angle OCA = \angle OAC = \beta$. Snell's law at point D implies $n\sin\beta = \sin\angle C_1CC_3$ and consequently $\angle C_1CC_3 = \alpha$. We further have $\angle C_1CC_2 = \angle COD = 180° - \alpha - \angle AOC$ and $\angle AOC = 180° - 2\beta$, which gives $\angle C_1CC_2 = 2\beta - \alpha$. We finally obtain $\theta_0 = \angle C_1CC_3 - \angle C_1CC_2 = 2\alpha - 2\beta = 2\alpha - 2\arcsin\left(\frac{\sin\alpha}{n}\right)$.

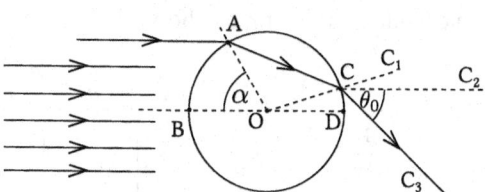

Figure 4.29 With the solution of problem 26(g)

The graph of the dependence of θ_0 on α for green light ($n = 1.333$) is shown in Figure 4.30. Since this dependence is monotonous, unlike the dependences in Figures 4.26 and 4.28, there is no characteristic direction of propagation of the majority of refracted rays. For this reason, these rays do not form a rainbow.

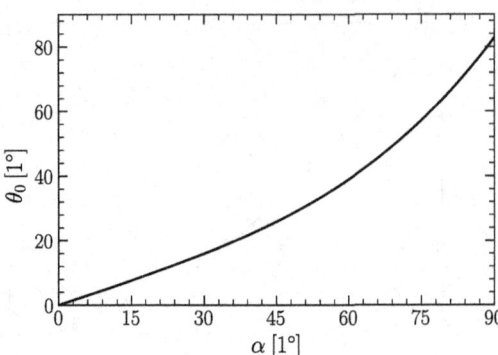

Figure 4.30 With the solution of problem 26(g)

We refer the reader interested in more details about the physics of rainbows to reference [24]. The authors presented a part of this problem at the competition for the selection of the Serbian team for the European Physics Olympiad in 2020.

Problem 27   Running and Driving in the Rain

How can people minimize getting wet in the rain?
How much rain falls on the windshield of a car?
How can a driver prevent rain falling on the rear windshield?

For most people it is unpleasant to stand in the rain. For this reason, people run to the closest shelter when it starts to rain. We investigate in this problem how much a person gets wet when it starts to rain and what should be done to get wet as little as possible. We assume that the shape of the person is a cuboid of height $h = 2.0\,\text{m}$ whose base is a square of side $a = 20\,\text{cm}$, as shown in Figure 4.31.

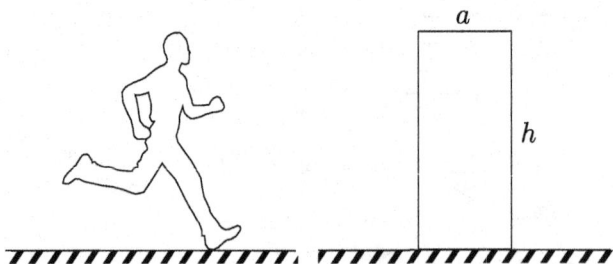

Figure 4.31 A person running in the rain and its simplified model

On a rainy day the volume of water that falls in unit time on unit area of the ground is $q = 0.50\,\frac{\text{cm}}{\text{h}}$. Raindrops fall vertically downward and their speed is $v = 4.0\,\frac{\text{m}}{\text{s}}$.

(a) The volume fraction of raindrops is defined as the ratio of the total volume of all raindrops in a certain part of space and the volume of that part of space. Determine the volume fraction of raindrops.

(b) What is the volume of water that falls on a person in $t = 30\,\text{min}$ if the person is standing?

(c) What is the volume of water that falls in $t = 30\,\text{min}$ on a person who runs at a speed of $u = 2.0\,\frac{\text{m}}{\text{s}}$? The direction of running is perpendicular to the side face of the cuboid.

(d) In the moment when it started to rain, a person was in a meadow at a distance $L = 1.0\,\text{km}$ from the closest shelter. The person wants to get wet as little as possible – that is, to minimize the amount of water that will fall on him/her on the way to the shelter. What should be his/her running speed?

On a rainy day it is more pleasant to drive a car than to run. We consider a car shown in Figure 4.32 whose front and rear windshields of length $l = 1.0\,\text{m}$ are positioned at an angle $\beta = 45°$ with respect to the horizontal and whose width is $b = 2.0\,\text{m}$. Weather conditions are the same as in previous parts of the problem.

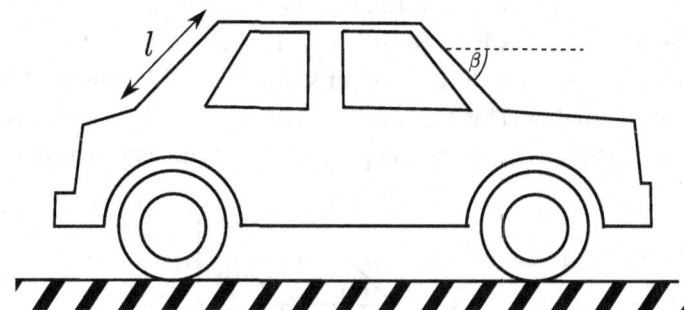

Figure 4.32 A car: the length of front and rear windshield is l. They are positioned at an angle β with respect to the horizontal.

(e) What is the minimal car speed $u_{\min}$ that prevents the rain from falling on the rear windshield?

(f) Show that for a speed smaller than $u_{\min}$ the volume of water falling on a windshield in unit time V' does not depend on the speed of the car. Calculate the numerical value of V'.

(g) Determine the dependence of the volume of water falling on windshields in unit time on the speed of the car u when $u > u_{\min}$.

Solution of Problem 27

(a) In time t the water of volume $V_y = qSt$ falls on the area S of the ground. During the same time all raindrops that were initially inside the volume $V_z = Svt$ fall on the same surface. Therefore, we find that the volume fraction of water is $r = \frac{V_y}{V_z} = \frac{q}{v} = 3.47 \cdot 10^{-7}$.

(b) The volume of water that falls on the person is $V_a = qta^2$, which leads to $V_a = 0.10 \, \text{l}$.

(c) We consider the phenomenon in the reference frame of the person. The person is at rest in this reference frame, while the velocity of the raindrops is $\vec{v}_r = \vec{v} - \vec{u}$ and it is directed at an angle α with respect to the horizontal (Figure 4.33). In time t all raindrops that were initially inside the parallelepiped P_1 (shown in Figure 4.33) fall on the side face of the cuboid. The base of the parallelepiped P_1 is the parallelogram of sides h and $v_r t$ and the height of the parallelepiped is a. In the same time t all raindrops that were initially inside the parallelepiped P_2 (shown in Figure 4.33) fall on the top face of the cuboid. The base of the parallelepiped P_2 is the parallelogram of sides a and $v_r t$, while its height is a. The volume of P_1 is $V_1 = v_r t a h \cos \alpha$, while the volume of P_2 is $V_2 = v_r t a^2 \sin \alpha$. The volume of the water in question is $V_b = r(V_1 + V_2)$. Using the relations $v_r \cos \alpha = u$ and $v_r \sin \alpha = v$ we get $V_b = qta\left(a + h\frac{u}{v}\right)$, which leads to $V_b = 0.60 \, \text{l}$.

(d) From the solution of part (c), it follows that the volume of water that falls on the person in unit time is $V' = qa\left(a + h\frac{u}{v}\right)$. The time it takes the person to reach the shelter is $t = \frac{L}{u}$. Therefore, water of volume $V = V't$ falls on the person on the way to the shelter. This leads to $V = qaL\left(\frac{a}{u} + \frac{h}{v}\right)$. V is minimal when u is as large as possible. Therefore, the person should run as fast as possible.

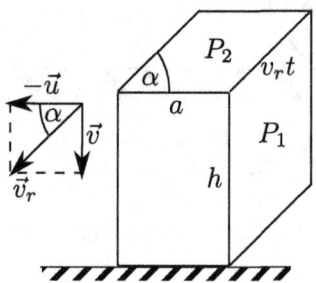

Figure 4.33 With the solution of problem 27(c)

(e) We consider the phenomenon in the reference frame of the car. The car is at rest in this frame, while the velocity of the drops is $\vec{v}_r = \vec{v} - \vec{u}$ and it is directed at angle α with respect to the horizontal (Figure 4.33). Raindrops will not fall on the rear windshield if $\alpha < \beta$. Using Figure 4.33 we conclude that the angle α satisfies $\tan \alpha = \frac{v}{u}$. Consequently the raindrops do not fall on the rear windshield when $u > \frac{v}{\tan \beta}$, i.e. $u_{min} = \frac{v}{\tan \beta} = 4.0 \, \frac{\text{m}}{\text{s}}$.

(f) In the case $u < u_{min}$, all raindrops that were initially in the parallelepiped P_1 (shown in Figure 4.34) fall on the rear windshield in time t. The base of P_1 is a parallelogram with sides l and $v_r \Delta t$, while its height is b. All raindrops that were initially in P_2 (shown in Figure 4.34) fall on the front windshield in time t. The base of P_2 is a parallelogram with sides l and $v_r \Delta t$, while its height is b. The volume of P_1 is $\Delta V_1 = v_r \Delta t l b \sin(\alpha - \beta)$, while the volume of P_2 is $\Delta V_2 = v_r \Delta t l b \sin(\alpha + \beta)$. The volume of water that falls on windshields in unit time is $V' = r\left(\frac{\Delta V_1}{\Delta t} + \frac{\Delta V_2}{\Delta t}\right)$. Using the identity $\sin(\alpha - \beta) + \sin(\alpha + \beta) = 2\sin\alpha\cos\beta$ and the relation $v_r \sin\alpha = v$ we obtain $V' = 2qlb\cos\beta = 1.4 \cdot 10^{-2} \frac{m^3}{h}$.

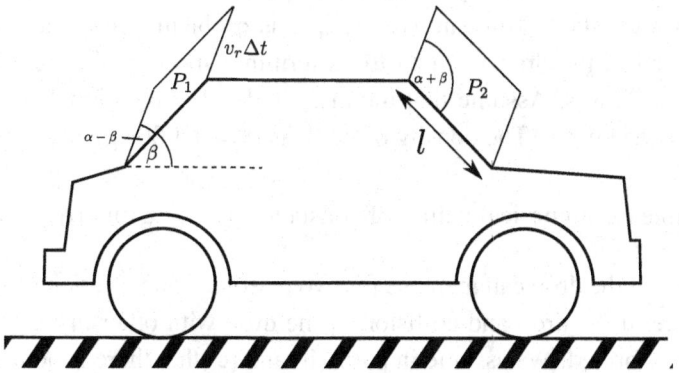

Figure 4.34   With the solution of problem 27(f)

(g) In this case all raindrops that were initially in P_2 fall on the front windshield in time Δt. The volume of water that falls on the front windshield in unit time is $V' = r\frac{\Delta V_2}{\Delta t}$. From the identity $\sin(\alpha + \beta) = \sin\alpha\cos\beta + \sin\beta\cos\alpha$ and the relations $v_r \cos\alpha = u$ and $v_r \sin\alpha = v$ we get $V' = qlb\left(\cos\beta + \frac{u}{v}\sin\beta\right)$.

The authors presented part of this problem at a regional physics competition for the fourth grade of high school in Serbia in 2016.

Problem 28   Rain

How do waterdrops form and grow?
How much time is needed for a waterdrop to grow?
How large should a waterdrop be in order to reach the ground before it evaporates?

Raindrops are formed by condensation of water vapor in the atmosphere. In this problem we consider the formation of raindrops, the time necessary for that, and their motion in the atmosphere, as well as how large the raindrop should be in order to reach the surface of the Earth before it evaporates. Assume in all parts of the problem that the shape of the raindrop is spherical and neglect the capillary effects and the buoyant force in the air.

Small raindrops move slowly through the atmosphere so that the flow of air around the drop is laminar. The terminal speed of the drop is then $v = c_1 R^{\alpha_1}$, where R is the radius of the drop, while c_1 and α_1 are constant coefficients. The flow is laminar for drops that have a radius of $R \leq R_1 = 70\,\mu\mathrm{m}$. Air flow around big drops with radius $R \geq R_2 = 1.0\,\mathrm{mm}$ is turbulent. The dependence of the terminal speed on the radius is then $v = c_3 \sqrt{R}$, where c_3 is a constant. In the intermediate regime for drops with radius $R_1 \leq R \leq R_2$ the terminal speed of the drop is given as $v = c_2 R$, where c_2 is a constant. You can assume in this problem that at each moment of time the speed of the drop is equal to its terminal speed. The viscosity of air is $\eta = 1.82 \cdot 10^{-5}\,\mathrm{Pa \cdot s}$. Assume that the magnitude of gravitational acceleration is constant $g = 9.81\,\mathrm{m/s^2}$. The density of water is $\rho_l = 1{,}000\,\mathrm{kg/m^3}$.

(a) Determine the numerical values of constants c_1, c_2, c_3, and α_1.

The radius of the drop can increase for two reasons: condensation of water vapor at the surface of the drop and collision of the drop with other drops when a new, larger drop is formed. We assume in parts (b) and (c) that there is no condensation and that the radius of the drop increases solely due to collisions with other drops. We consider a drop with a radius in the range $R_1 \leq R \leq R_2$. Assume that all the drops it collides with are much smaller, that it fully absorbs each drop it collides with, that small drops are uniformly distributed in space, and that their density is $\rho_b = 2.0 \cdot 10^{-3}\,\mathrm{kg/m^3}$.

(b) Determine the time dependence of the radius of this drop. Its radius at time $t = 0$ is R_0.

(c) Calculate the time it takes for this drop to grow from $R_1 = 70\,\mu\mathrm{m}$ to $R_2 = 1.0\,\mathrm{mm}$.

In parts (d)–(i) we consider a drop that does not collide with other drops. Its radius changes only due to evaporation of water from its surface and condensation of surrounding water vapor. In parts (d)–(g) you can assume that the center of mass of the drop is at rest. In the course of condensation the water vapor in the layer outside the drop turns into a liquid and hence becomes a part of the drop. The condensation process is accompanied by the diffusion of the vapor outside the drop, which leads to redistribution of the vapor density near the drop. The diffusion law states that the mass of water vapor that in unit time passes through the sphere of radius r concentric with the drop (this sphere is shown in Figure 4.35) is

$$q_m = -DS\frac{d\rho_v(r)}{dr}, \tag{4.38}$$

where $D = 2.2 \cdot 10^{-5}\,\mathrm{m^2 s^{-1}}$ is the water vapor diffusion coefficient, ρ_v is its density, and $S = 4r^2\pi$ is the area of the sphere of radius r. The assumed direction of particle flow in this equation is toward the outside of the sphere. Assume that the radius of the drop changes slowly so that the flow of water vapor particles is steady.

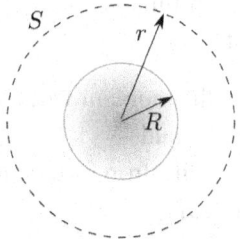

Figure 4.35 Raindrop of radius R and concentric sphere of radius r

(d) Derive and expression that links the quantities $q_m(r)$, $\rho_v(r)$, and $\rho_v(\infty)$.

The law of heat conduction states that the amount of heat that in unit time passes through the sphere of radius r concentric with the drop (this sphere is shown in Figure 4.35) is

$$q_q = -KS\frac{dT(r)}{dr}, \tag{4.39}$$

where $K = 2.5 \cdot 10^{-2}\,\mathrm{Wm^{-1}K^{-1}}$ is the heat conductivity of water vapor, while $T(r)$ is the temperature at the distance r from the center of the drop. The assumed direction of heat flow in this equation is toward the outside of the drop. Assume that the radius of the drop changes slowly so that the heat flow is steady.

The specific heat of water evaporation is $L = 2.5 \cdot 10^6\,\mathrm{J/kg}$, the molar mass of water is $M_v = 18\,\mathrm{g/mol}$, and the gas constant is $R_g = 8.31\,\mathrm{J \cdot mol^{-1}K^{-1}}$. The temperature at a sufficiently large distance from the drop is $T_\infty = 280\,\mathrm{K}$, while the relative humidity is $r_H = 101\,\%$. Relative humidity is defined as the ratio of water vapor pressure and the saturated water vapor pressure at the same temperature. Assume that water vapor is an ideal gas and that the saturated water vapor pressure in the relevant range of temperatures is given as

$$p_s = [a_1 + a_2\,(T - T_\infty)] \cdot T, \tag{4.40}$$

where $a_1 = 3.571\,\mathrm{Pa/K}$ and $a_2 = 0.333\,\mathrm{Pa/K^2}$.

(e) Determine the change of the drop mass in unit time at the moment when the drop radius is R.

(f) Determine the time dependence of the drop radius if its radius at time $t = 0$ is R_0.

(g) Calculate the time it takes for the drop to grow from $R_1 = 70\,\mu m$ to $R_2 = 1.0\,mm$.

The drop formed in the cloud at the height of $H = 1.0\,km$ above the ground is falling through the atmosphere. The relative humidity is $r'_H = 70\,\%$. Assume that the change of the drop mass in unit time is given by the same law (derived in part [e]) as for the drop that is at rest.

(h) Determine the drop radius at the moment when it is at height h if its radius was R_0 when it was formed in the cloud.

(i) For what value of the radius of the drop formed in the cloud does the drop reach the ground before it evaporates?

Solution of Problem 28

(a) The drag force in the laminar regime is given by the Stokes law as $F = 6\pi\eta R v$, where $v = c_1 R_1^\alpha$. When the drop reaches terminal speed the drag force is in equilibrium with the gravity force. Consequently $F = mg$, where $m = \frac{4}{3}\rho_l R^3 \pi$ is the mass of the drop. From previous equations we obtain $\alpha_1 = 2$ and $c_1 = \frac{2\rho_l g}{9\eta} = 1.2\cdot 10^8\,s^{-1}m^{-1}$. From two different expressions for the speed at $R = R_1$, it follows that $c_1 R_1^2 = c_2 R_1$ and consequently $c_2 = c_1 R_1 = 8.4\cdot 10^3\,s^{-1}$. From expressions for the speed at $R = R_2$, we obtain $c_2 R_2 = c_3\sqrt{R_2}$ and therefore $c_3 = c_2\sqrt{R_2} = 265\,m^{1/2}s^{-1}$.

(b) In time dt the drop absorbs all the drops inside the volume $dV = R^2\pi v dt$. Therefore, the mass change during this time is $dm_k = \rho_b R^2\pi v dt$. We also have $dm_k = \rho_l 4R^2\pi dR$ and $v = c_2 R$. From the previous three equations we obtain $\frac{dR}{R} = \frac{c_2\rho_b}{4\rho_l}dt$. After performing the integration, it follows that $R(t) = R_0\exp\left(\frac{c_2\rho_b}{4\rho_l}t\right)$.

(c) From the solution of part (b) we obtain $t = \frac{4\rho_l}{c_2\rho_b}\ln\frac{R_2}{R_1} = 10.6\,min$.

(d) Since the system is in steady state, the mass conservation law leads to $q_m(r) = $ const. The diffusion law implies $q_m = -D\cdot 4r^2\pi\frac{d\rho_v(r)}{dr}$ and therefore $d\rho_v(r) = -\frac{dr}{r^2}\frac{q_m}{4\pi D}$. After performing the integration of previous equation from a certain r to $r = \infty$ one obtains $\rho_v(\infty) - \rho_v(r) = -\frac{q_m}{4\pi Dr}$.

(e) Applying the solution of part (d) for $r = R$ we have

$$q_m(R) = -4\pi DR[\rho_v(\infty) - \rho_v(R)]. \tag{4.41}$$

In steady state we have $q_q(r) = \text{const.} = q_q(R)$. The heat conduction law gives $q_q = -K \cdot 4r^2 \pi \frac{dT(r)}{dr}$ and consequently $dT(r) = -\frac{dr}{r^2} \frac{q_q}{4\pi K}$. After performing the integration from $r = R$ to $r = \infty$ we find

$$q_q(R) = -4\pi K R [T_\infty - T(R)]. \qquad (4.42)$$

During the time dt when the mass dm of vapour condenses, the amount of heat released is Ldm. Therefore, $q_q(R) = L\frac{dm}{dt}$ and consequently

$$q_q(R) = -Lq_m(R). \qquad (4.43)$$

Relative humidity is

$$r_H = \frac{p_v}{p_s(T_\infty)}, \qquad (4.44)$$

where p_v is the vapor pressure at a large distance from the drop. The equation of the state of the ideal gas at a large distance from the drop reads as

$$p_v = \frac{\rho_v(\infty) R_g T_\infty}{M_v}. \qquad (4.45)$$

Liquid water and water vapor are in equilibrium at the surface of the drop. Therefore, the pressure of water vapor is equal to saturated water vapor pressure and satisfies the ideal gas equation

$$p_s[T(R)] = \frac{\rho_v(R) R_g T(R)}{M_v}. \qquad (4.46)$$

The dependence of the saturated water vapor pressure on temperature given in the text of the problem implies

$$p_s(T_\infty) = a_1 T_\infty \qquad (4.47)$$

and

$$\frac{p_s[T(R)]}{T(R)} = a_1 + a_2[T(R) - T_\infty]. \qquad (4.48)$$

From equations (4.41)–(4.48) one obtains

$$q_m = -aR, \qquad (4.49)$$

where

$$a = \frac{4\pi K M_v a_1}{L R_g} \frac{D(r_H - 1)}{\frac{K}{L} + Da_2 \frac{M_v}{R_g}} = 8.3 \cdot 10^{-9} \frac{\text{kg}}{\text{m} \cdot \text{s}}. \qquad (4.50)$$

The change of the drop mass in unit time is finally $\frac{dm_k}{dt} = -q_m = aR$.

(f) Since $\frac{dm_k}{dt} = -q_m$ and $dm_k = \rho_l 4R^2 \pi dR$, it follows that

$$d(R^2) = \frac{a}{2\pi\rho_l} dt. \tag{4.51}$$

By performing the integration from 0 to t we obtain $R(t) = \sqrt{R_0^2 + \frac{a}{2\pi\rho_l} t}$.

(g) Using the solution of part (f) we obtain $t = \frac{2\pi\rho_l}{a}\left(R_2^2 - R_1^2\right) = 8.75$ days.

(h) From $v = c_2 R$, $\frac{dh}{dt} = -v$ and equation (4.51) it follows that $R^2 dR = \frac{a'}{4\pi\rho_l c_2} dh$, where $a' = \frac{4\pi K M_v a_1}{LR_g} \frac{D(1-r'_H)}{\frac{K}{L}+Da_2\frac{M_v}{R_g}}$. After integration we find $R(h) = \sqrt[3]{R_0^3 - \frac{3a'}{4\pi\rho_l c_2}(H-h)}$.

(i) The drop reaches the ground when the condition $R(h=0) > 0$ is satisfied. The solution of part (h) leads to $R_0 > \sqrt[3]{\frac{3a'H}{4\pi\rho_l c_2}} = 0.19$ mm.

We refer the reader interested in more details about the formation of raindrops to reference [2]. The authors presented this problem at the Serbian Physics Olympiad for high school students in 2016.

Problem 29   Lightning

What is lightning and what is thunder?
How many lightning strikes happen on a daily basis?

In this problem we consider the circulation of charge in the Earth's atmosphere including lightning that is followed by thunder. The ionosphere is the upper layer of the atmosphere that is positively charged due to the influence of cosmic radiation. The surface of the Earth is negatively charged. Consequently the surface of the Earth and the ionosphere form a large natural capacitor known as an atmospheric capacitor. When the weather is nice, the magnitude of electric field just above the surface of the Earth is $E_0 = 100 \frac{V}{m}$. This field is directed towards the center of the Earth.

You can assume in all parts of the problem that the Earth is a conductor in the shape of a ball with radius $R = 6,400$ km and that the ionosphere is at a height $H = 80$ km above the surface of the Earth (Figure 4.36). Dielectric permittivity of air is $\varepsilon_0 = 8.85 \cdot 10^{-12} \frac{F}{m}$.

(a) Calculate the amount of charge on the surface of the Earth.
(b) Calculate the potential difference between the Earth and the ionosphere and the capacitance of the atmospheric capacitor. Assume in this part of the problem that air is a perfect insulator.

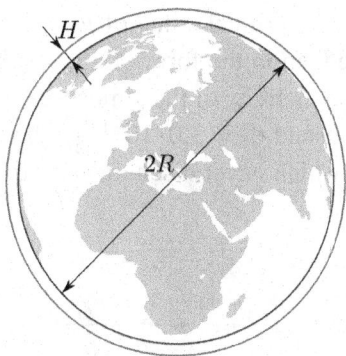

Figure 4.36 Atmospheric capacitor

(c) Using the fact that the height of the ionosphere is much smaller than the Earth's radius show that the atmospheric capacitor can be considered as a parallel-plate capacitor.

Air contains a low concentration of ions, which leads to significant differences of the electric field and conductance in the atmosphere from those predicted by the model considered so far. In reality, air is not an ideal insulator, but it is a weak conductor. The specific conductance of atmosphere at a height h is given as

$$\sigma(h) = \sigma_0 \, e^{h/\Delta}, \qquad\qquad (4.52)$$

where $\sigma_0 = 2.0 \cdot 10^{-13} \frac{\text{S}}{\text{m}}$ is the conductance of air at the Earth's surface, while $\Delta = 5.0\,\text{km}$. Positively charged particles move toward Earth's surface, while negatively charged particles move toward the ionosphere. Such directed motion of particles is the current that discharges the atmospheric capacitor.

(d) Calculate the electric resistance of the atmosphere and the capacitance of the atmospheric capacitor.
(e) Estimate the time of the capacitor discharging, defined as the time it takes for the charge of the capacitor to decrease to 0.1% of its initial value.

Although it is expected that the atmospheric capacitor should discharge in a couple of minutes only, this does not happen due to thunderstorms. Thunderstorm clouds, so-called cumulonimbus clouds, have an average area of $S = 100\,\text{km}^2$ and form at an average height of $H_1 = 3.0\,\text{km}$. Lower parts of cumulonimbus clouds are negatively charged. This induces a positive charge at high objects on the Earth (tree, tower, building). Consequently a homogeneous electric field is formed between the clouds and the Earth. This field has an opposite direction to the direction of the electric field in the atmospheric capacitor. Due to the accumulation of charge in the region where thunderstorm clouds are formed the magnitude of this electric field increases and reaches the critical value of $E_1 = 30\ \frac{\text{kV}}{\text{m}}$. Electric discharge in the atmosphere takes place then in the form of a giant spark. Lightning is formed (Figure 4.37), while the conductivity of the atmosphere between the cloud and the Earth's surface becomes constant $\sigma_1 = 3.0 \cdot 10^{-8}\ \frac{\text{S}}{\text{m}}$. This contributes to charging of the atmospheric capacitor, which never discharges for this reason.

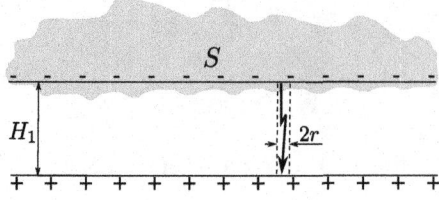

Figure 4.37 Lightning

(f) Estimate the time interval of electrical discharge of the lightning and the number of lightning strikes that happen every day.

We consider next what happens to the air during the passage of the lightning. We assume that lightning passes through a cylinder of radius $r = 11$ cm. Assume that the complete electrical energy during the passage of the lightning transforms to heat, which strongly heats the air. Sudden heating explosively spreads the air and a bang is heard – the thunder. The thunder can be heard at different but closely spaced moments of time during the passage of the lightning. For this reason, a continuous sound wave is heard – that is, the thunder. The heat capacity of air is $C = 1,000\ \frac{\text{J}}{\text{kgK}}$, while the density of air is $\rho = 1.0\ \frac{\text{kg}}{\text{m}^3}$. The speed of sound in air v depends on the temperature of air T as $v = \sqrt{\alpha T}$, where $\alpha = 400\ \frac{\text{m}^2}{\text{Ks}^2}$.

(g) Calculate the temperature T_1 of the heated air in the cylinder due to the passage of the lightning. The temperature of air was $T_0 = 27°C$ before the formation of lightning.
(h) Why is lightning seen before thunder is heard?
(i) Calculate the time interval between the moment when an observer on the Earth sees lightning and the moment when the observer hears thunder. The sound wave of the thunder that reaches the observer travels $l = 4.0$ km. The temperature of air at the position of the observer is T_0. The temperature of air decreases exponentially away from the lightning as

$$T(x) = A e^{-Bx}, \qquad (4.53)$$

where x is the distance from the lightning, while A and B are positive constants.

Solution of Problem 29

(a) Since the Earth is an ideal conductor, a negative charge $-Q$ is distributed across its surface. Using Gauss's law $E \cdot S = \frac{-Q}{\varepsilon_0}$ for the spherical surface $S = 4r^2\pi$ of radius r $(R \leq r \leq R+H)$, we obtain the magnitude of electric field on this surface $E(r) = \frac{1}{4\pi\varepsilon_0}\frac{Q}{r^2}$. The electric field at the surface of the Earth is $E(r = R) = E_0$. Consequently the amount of charge on the Earth is $Q = 4\pi\varepsilon_0 R^2 E_0 = 4.56 \cdot 10^5 \, \text{C}$.

(b) The potential difference between the Earth's surface and the ionosphere is

$$V_1 = \int_R^{R+H} E(r)dr = \frac{Q}{4\pi\varepsilon_0}\left(\frac{1}{R} - \frac{1}{R+H}\right) = E_0\frac{RH}{R+H} = 7.90 \cdot 10^6 \, \text{V}, \quad (4.54)$$

while the capacitance of atmospheric capacitor is $C_1 = \frac{Q}{V_1} = 4\pi\varepsilon_0\frac{R(R+H)}{H} = 58 \, \text{mF}$.

(c) Using the condition $H \ll R$, equations for the voltage between the electrodes and the capacitance of the capacitor reduce to $V_1 = E_0 H$ and $C_1 = \varepsilon_0\frac{4\pi R^2}{H}$, which are equivalent to the equations for a parallel-plate capacitor with plates of area $4\pi R^2$ at a mutual distance of H.

(d) The electrical resistance of the atmosphere is

$$\mathcal{R} = \int_0^H \frac{1}{\sigma(h)}\frac{dh}{4R^2\pi} = -\frac{\Delta}{4\pi R^2\sigma_0}\left(e^{-\frac{H}{\Delta}} - 1\right) \approx \frac{\Delta}{4\pi R^2\sigma_0} = 48.6\,\Omega. \quad (4.55)$$

Due to the conductivity of the atmosphere, the current flows through it and discharges the atmospheric capacitor. The current density is constant at all heights, which follows from the continuity equation $j_1 4\pi R_1^2 = j_2 4\pi R_2^2$, which leads to $j_1 \approx j_2$ since $R_1 \approx R_2$. It follows from Ohm's law $j = \sigma E = const$ that the electric field decreases exponentially as the height increases as $E(h) = \frac{j}{\sigma} = E_0 e^{-h/\Delta}$, where we used the fact that the magnitude of electric field at the surface of the Earth is $E_0 = j/\sigma_0$. The difference of potentials between the Earth's surface and the ionosphere is

$$V_2 = \int_0^H E(h)dh = -E_0\Delta\left(e^{-\frac{H}{\Delta}} - 1\right) \approx E_0\Delta, \quad (4.56)$$

while the capacitance of the atmospheric capacitor is $C_2 = \frac{Q}{V_2} = 4\pi\varepsilon_0\frac{R^2}{\Delta} = 0.91 \, \text{F}$.

(e) The capacitor C_2 discharges through the resistor $\mathcal{R}$ where it holds that $U_{C_2} + U_{\mathcal{R}} = 0$, which leads to $\frac{q}{C_2} + \frac{dq}{dt}\mathcal{R} = 0$. The last equation is a differential equation with separated variables

$$\int \frac{dq}{q} = -\int \frac{dt}{C_2\mathcal{R}}. \quad (4.57)$$

After integration, it follows that $\ln q = -\frac{t}{C_2 \mathcal{R}} + K_1$, where K_1 is the integration constant. K_1 is determined from the initial condition $q(t=0) = Q$, which leads to $K_1 = \ln Q$ and $q(t) = Q e^{-t/\tau_1}$, where $\tau_1 = C_2 \mathcal{R} = \varepsilon_0/\sigma_0$ is the so-called time constant. The discharge time of the capacitor can be estimated by calculating the time it takes for the charge of the capacitor to decrease $1{,}000$ times, which leads to $t_p = \tau_1 \ln(1{,}000) = 306\,\text{s}$.

(f) The surface of the Earth below the cloud and the lower surface of the cloud form a capacitor of capacitance $C_3 = \varepsilon_0 S/H_1$, while the resistance of the air that fills this capacitor is $\mathcal{R}_1 = H_1/(\sigma_1 S)$. Therefore, the capacitor C_3 discharges via a resistor $\mathcal{R}_1$ where it holds that $U_C + U_{\mathcal{R}_1} = 0$. In a similar manner as in part (e) it follows that $\ln q = -\frac{t}{C_3 \mathcal{R}_1} + K_2$, where K_2 is the integration constant that is determined from the initial condition $q(t=0) = q_0$, where $q_0 = U_C C_3 = \varepsilon_0 E_1 S$ is the amount of charge before the start of discharging. Consequently $K_2 = \ln q_0$, i.e. $q(t) = q_0 e^{-t/\tau_2}$, while $\tau_2 = C_3 \mathcal{R}_1 = \varepsilon_0/\sigma_1$. The duration of lightning is the time it takes for the capacitor to discharge - that is, the time it takes for the charge of the capacitor to decrease $1{,}000$ times, which leads to $t_m = \tau_2 \ln(1{,}000) = 2.0\,\text{ms}$. Since the atmospheric capacitor does not discharge due to lightning events, $n = Q/q_0$ lightning events happen during time t_p. Consequently, during the time $t_{day} = 24\,\text{h}$, the number of lightning events is $N_2 = t_{day} n/t_p \approx 4.85 \cdot 10^6$.

(g) The electrical energy of the lightning can be estimated as the energy of the capacitor formed by the Earth's surface and the lower surface of the clouds, which leads to $\mathcal{E} = \frac{1}{2} q_0 U = \frac{1}{2} \varepsilon_0 E_1^2 H_1 S$. The complete energy $\mathcal{E}$ is used to heat an air column of height H_1 and cross section area $r^2 \pi$, which leads to $\mathcal{E} = \rho r^2 \pi H_1 C(T_1 - T_0)$ and $T_1 = T_0 + \frac{\varepsilon_0 E_1^2 S}{2\rho r^2 \pi C} = 1.08 \cdot 10^4\,\text{K}$.

(h) The speed of sound is much slower than the speed of light. For this reason, the time it takes for the sound to come to the observer is much longer than the time it takes for the light to reach the observer. For this reason, lightning is seen before thunder is heard.

(i) The constants A and B can be found from the conditions $T(x=0) = T_1$ and $T(x=l) = T_0$, which leads to $A = T_1$ and $B = -\frac{1}{l} \ln\left(\frac{T_0}{T_1}\right)$. The equation $T(x) = A e^{-Bx}$ then becomes $T(x) = T_1 \left(\frac{T_0}{T_1}\right)^{x/l}$. It follows from the dependence of the speed of sound in air on temperature $v = \sqrt{\alpha T}$ that $v = \frac{dx}{dt} = \sqrt{\alpha T_1} \left(\frac{T_0}{T_1}\right)^{\frac{x}{2l}}$. This is a differential equation with separated variables $dt = \frac{1}{\sqrt{\alpha T_1}} \left(\frac{T_1}{T_0}\right)^{\frac{x}{2l}} dx$. By performing the integration, we obtain the time t_g when the observer hears the thunder $t_g = \dfrac{2l\left(\sqrt{\frac{T_1}{T_0}} - 1\right)}{\sqrt{\alpha T_1} \ln\left(\frac{T_1}{T_0}\right)} = 5.37\,\text{s}$. In the same time, this is the time interval between the moment when the observer on the Earth sees the lightning and the moment when the observer hears the thunder.

We refer interested readers to reference [32]. The authors presented this problem at the Serbian Physics Olympiad for high school students in 2016.

Problem 30   Ocean Waves

How do ocean waves form?
What is the difference between the waves in deep and shallow water?

Ocean waves are formed as disturbances caused by wind, by the motion of big objects (such as coastal landslides, icebergs, ocean floor), and by the gravitational attraction of the moon and the Earth, as well as the sun and the Earth. The energy released by the source of the wave causes water particles to move on a circular orbit, transferring energy between the water particles. This way an ocean wave is formed at the boundary of water and air. The circular motion of water particles on the surface of the ocean continues also under water, as shown in Figure 4.38. The diameter of the orbit of particles of water decreases as depth increases, so that the water at the bottom of the ocean is at rest. Most characteristics of ocean waves depend on the ratio of the wavelength and the depth of the water. The wavelength determines the size of the orbits of the water particles in the wave, while the depth of water determines the shape of the orbits.

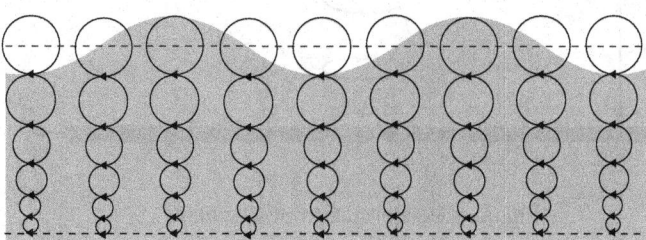

Figure 4.38 Formation of a wave due to local motion of water particles

(a) Lamb showed that the connection between the angular frequency of the wave ω and wave number k (dispersion relation) is of the form:

$$\omega^2 = gk\tanh(kd), \qquad\qquad (4.58)$$

where g is gravitational acceleration and d is the depth of the ocean. Calculate the speed of propagation of ocean waves in cases when the depth of the ocean is much shallower and much deeper than the wavelength.

(b) Wave 1 propagates in water whose depth is deeper than half a wavelength (deep water), while wave 2 propagates in water whose depth is shallower than $1/20$ of the wavelength (shallow water). Determine the expression for speed of propagation of waves 1 and 2.

Big water waves of huge energy and devastating power, called tsunamis, can form as a consequence of underwater earthquakes, explosions, and volcanic eruptions, and less frequently due to nuclear tests or meteor impacts. The most devastating tsunamis from recent history were caused by the Sumatra–Andaman earthquake in 2004 and the Tohoku earthquake in 2011. A tsunami is formed when particles of water move locally on elliptical trajectories at an angular velocity ω, as shown in Figure 4.39. The elongation Δx with respect to the position (x, y) when there is no wave, depends on time as:

$$\Delta x = \Delta x_0 \sin(kx - \omega t). \tag{4.59}$$

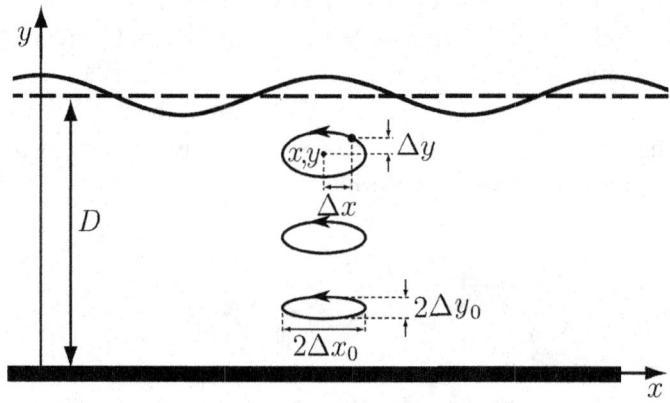

Figure 4.39 Formation of a tsunami

(c) The length of the half axis in the y-direction linearly increases with y, where the corresponding proportionality coefficient is $\Delta x_0 k$. Determine the dependence of elongation Δy on time. Determine the equation of the tsunami $\Delta y(x,t)$ at the surface of the ocean. The depth of the ocean when there is no wave is D.
(d) Determine the speed of the water particles that form the tsunami.

To find the dispersion relation of the tsunami, we consider a standing wave obtained as a linear combination of two identical tsunamis that move in opposite directions along the x-axis, described as:

$$\Delta x_+ = \Delta x_0 \sin(kx - \omega t), \tag{4.60}$$
$$\Delta x_- = \Delta x_0 \sin(kx + \omega t). \tag{4.61}$$

Neglect the viscosity and compressibility of water.

(e) Determine the time dependence of the elongations Δx and Δy in this case. Determine the moments of time when the surface of the ocean is flat and the moments of time when it is maximally deformed.

(f) The wavelength of the tsunami is much longer than the depth of the ocean. Prove that the dispersion relation for the tsunami is the same as the dispersion relation for the waves in shallow water.

The epicenter of the Sumatra-Andaman earthquake, which took place at $t_0 = 7:58$ local time, was at a distance of $s_1 = 160\,\text{km}$ from the island of Sumatra. The tsunami caused by this earthquake arrived at the island at $t_1 = 8:18$. It also propagated east via the Andaman basin and sea and reached the coasts of Thailand and Myanmar at $t_2 = 9:07$. Propagating west via the Indian Ocean, it reached Sri Lanka at $t_3 = 10:10$. This tsunami was felt globally. It reached the east coast of Africa after 7 h, it arrived in Rio de Janerio via the Atlantic Ocean after 20 h, it reached Chile after 23 h via the Pacific Ocean, and it even reached Halifax in Canada after 29 h.

(g) What is the depth of the ocean d_1 at the epicenter of the earthquake? Assume that the depth of water linearly decreases from the epicenter to Sumatra. Gravitational acceleration is $g = 9.83\,\text{m/s}^2$.

(h) The average depth of the Andaman Sea is $d_2 = 1.00\,\text{km}$, while the average depth of the Indian Ocean is $d_3 = 5.00\,\text{km}$. What are the speeds of the tsunamis that traveled to Thailand and Sri Lanka and what distances did these tsunamis travel before reaching the respective coasts?

(i) The height of the antinode of the wave in the region near the epicenter was $\Delta y_{01} = 40.0\,\text{cm}$. The waves generated by wind are of similar size. It is therefore difficult to distinguish between these two types of waves far away from the coast. However, unlike the waves generated by wind, the height of the antinode of the tsunami suddenly increases as the wave approaches the coast. This becomes visible in the last minute only, which makes it very difficult to detect these waves early enough. When the height of the antinode is $\Delta y_{02} = 2.00\,\text{m}$, this is a good indication that this is a tsunami. What is the depth of water at this place in this moment? Is it possible to escape from the tsunami if you are in water at that time?

Note: The function $\tanh x$ is defined as $\tanh x = \frac{e^x - e^{-x}}{e^x + e^{-x}}$.

Solution of Problem 30

(a) In the case when $d \gg \lambda = 2\pi/k$ - that is, when $kd \gg 1$, it holds that $\tanh(kd) \approx 1$, because $e^{-kd} \approx 0$, which leads to $\frac{e^{kd}-e^{-kd}}{e^{kd}+e^{-kd}} \approx \frac{e^{kd}}{e^{kd}} = 1$. The dispersion relation takes the form $\omega^2 = gk$, which leads to the speed of the wave of

$$v = \omega/k = \sqrt{g/k}. \tag{4.62}$$

In the case when $d \ll \lambda$, i.e. $kd \ll 1$, it holds that $\tanh(kd) \approx kd$, because $e^{kd} \approx 1 + kd$ and $e^{-kd} \approx 1 - kd$, which leads to $\frac{e^{kd}-e^{-kd}}{e^{kd}+e^{-kd}} \approx kd$. The dispersion relation takes the form $\omega^2 = gdk^2$, which in this case leads to the speed of the wave of

$$v = \omega/k = \sqrt{gd}. \tag{4.63}$$

(b) Wave 1 that propagates in deep water satisfies $d > \lambda/2 = \pi/k$, i.e. $kd > \pi$. The solid line in Figure 4.40 is the graph of the function $y = \tanh x$, which implies that for $x > 3$ the approximation $\tanh x \approx 1$ holds. From the result of part (a) we conclude that the speed of the wave in deep water is determined by equation (4.62). Wave 2 that propagates in shallow water satisfies $d < \lambda/20$ - that is, $kd < \pi/10$. The dashed line in Figure 4.40 is the graph of the function $y = x$. It is clear that the approximation $\tanh x \approx x$ holds for $x < 0.4$. From the result of part (a) we conclude that the speed of the wave in shallow water is determined by equation (4.63).

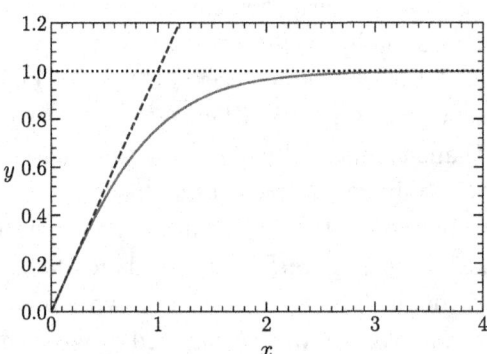

Figure 4.40 The graph of the function $y = \tanh x$ (solid line) and the graph of the function $y = x$ (dashed line)

The results for the dispersion relation for waves in deep and shallow water given by equations (4.62) and (4.63) can also be obtained by the techniques of "order of magnitude physics" with the help of dimensional analysis, which is described in more detail in reference [25].

(c) The trajectory of water particles is an ellipse described by the equation $\Delta x^2/\Delta x_0^2 + \Delta y^2/\Delta y_0^2 = 1$. The elongation in the y-direction is therefore $\Delta y = \pm \Delta y_0 \cos(kx - \omega t)$. The signs $\pm$ in the previous equation determine the direction of motion of the particles, where the $+$ sign corresponds to a positive mathematical direction and the $-$ sign to negative. We use in the following $\Delta y = \Delta y_0 \cos(kx - \omega t)$. The statement of the problem implies that $\Delta y_0 = \Delta x_0 ky$, which leads to $\Delta y = \Delta x_0 ky \cos(kx - \omega t)$. The equation of the tsunami at the surface of the ocean is $\Delta y = \Delta x_0 kD \cos(kx - \omega t)$.

(d) From the expression for elongations Δx and Δy, we calculate the corresponding components of the velocities $u_x = \frac{d(\Delta x)}{dt} = -\Delta x_0 \omega \cos(kx - \omega t)$ and $u_y = \frac{d(\Delta y)}{dt} = -\Delta x_0 ky \omega \sin(kx - \omega t)$. The speed of water particles is $u = \sqrt{u_x^2 + u_y^2} = \Delta x_0 \omega \sqrt{\cos^2(kx - \omega t) + k^2 y^2 \sin^2(kx - \omega t)}$.

(e) Elongations Δx and Δy of the standing wave formed by the superposition of two identical tsunamis described by equations (4.60) and (4.61) are

$$\Delta x = \Delta x_+ + \Delta x_- = \Delta x_0 \left[\sin(kx - \omega t) + \sin(kx + \omega t)\right] = 2\Delta x_0 \sin(kx) \cos(\omega t) \tag{4.64}$$

$$\Delta y = \Delta y_+ + \Delta y_- = \Delta x_0 ky \left[\cos(kx - \omega t) + \cos(kx + \omega t)\right]$$
$$= 2\Delta x_0 ky \cos(kx) \cos(\omega t) \tag{4.65}$$

From equation (4.65) it follows that the surface of the wave is described using the equation $\Delta y = 2\Delta x_0 kD \cos(kx) \cos(\omega t)$. The surface of the ocean is flat when $\Delta y = 0$ (Figure 4.41a) – that is, when $\cos(\omega t) = 0$, which leads to $t_n = (2n+1)\pi/(2\omega)$, where $n \in \mathbb{N}_0$. The surface of the ocean is maximally deformed when $\cos(\omega t) = \pm 1$, which leads to $t_n = n\pi/\omega$, where $n \in \mathbb{N}_0$.

(f) We determine the dispersion relation for the tsunami using the conservation of energy. We consider a part of the wave of width L and length λ shown in Figure 4.41b. When the surface of the ocean is flat, water particles have maximal speed and the total energy of the wave is equal to kinetic energy

$$E = E_k = \frac{1}{2} \int dm v^2 = \frac{\rho L}{2} \int_0^\lambda dx \int_0^D dy (v_x^2 + v_y^2), \tag{4.66}$$

$$E = 2(\Delta x_0)^2 \omega^2 \rho L \int_0^\lambda dx \int_0^D dy \left[\sin^2(kx) + k^2 y^2 \cos^2(kx)\right] \approx \frac{2\pi (\Delta x_0)^2 \omega^2 \rho LD}{k}. \tag{4.67}$$

The second integral in the previous equation was neglected because it holds that $\lambda \gg D$ – that is, $kD \ll 1$, which implies $v_y \ll v_x$. In the case when the surface of the ocean is maximally deformed, potential energy is equal to the total energy of the wave. This change of potential energy is equal to the work performed by moving the part of water from the region below to the region above sea level, as shown in Figure 4.41c. Consequently

$$E = E_p = A = \int_{\lambda/4}^{3\lambda/4} dmg\Delta y = 2\rho g L \int_{\lambda/4}^{\lambda/2} dx(\Delta y)^2$$

$$= 8\rho g L (\Delta x_0)^2 k^2 D^2 \int_{\lambda/4}^{\lambda/2} dx \cos^2(kx)$$

$$= 8\rho g L (\Delta x_0)^2 k^2 D^2 \int_{\lambda/4}^{\lambda/2} dx \cos^2(kx) = 2\pi \rho g L D^2 (\Delta x_0)^2 k.$$

$$(4.68)$$

Using equations (4.67) and (4.68) we obtain the dispersion relation of the tsunami $\omega^2 = gDk^2$.

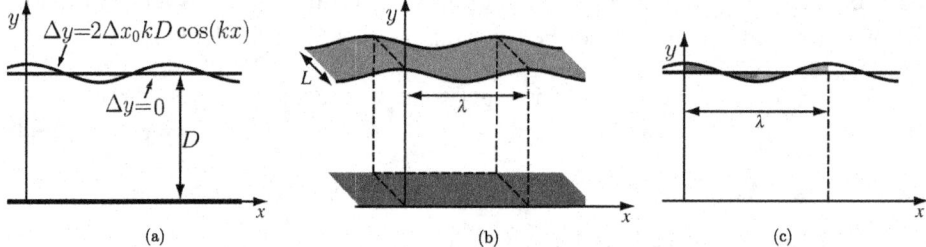

(a)   (b)   (c)

Figure 4.41 Maximally deformed ocean and volume element of the tsunami

(g) We denote as x the horizontal distance from the epicenter. On one hand, the speed of the tsunami is $v = dx/dt$. On the other hand, from the dispersion relation it follows that the speed of the tsunami is $v = \sqrt{gd(x)}$, where $d(x)$ is the depth at a given value of x. From the condition that the depth decreases linearly, it follows that $d(x) = d_1 \frac{s_1-x}{s_1}$. From previous equations, we obtain $dx/dt = \sqrt{\frac{gd_1(s_1-x)}{s_1}}$, which after the integration

$$\int_0^{s_1} \frac{dx}{\sqrt{s_1-x}} = \sqrt{\frac{gd_1}{s_1}} \int_0^{t_1-t_0} dt, \qquad (4.69)$$

gives $d_1 = \frac{4s_1^2}{g(t_1-t_0)^2} = 7.23\,\text{km}$.

(h) The tsunami was traveling at a speed $v_2 = \sqrt{gd_2} = 357\,\text{km/h}$ toward Thailand and it traveled a distance of $s_2 = v_2(t_2-t_0) = 410\,\text{km}$ before reaching the coast. It was traveling at a speed $v_3 = \sqrt{gd_3} = 798\,\text{km/h}$ toward Sri Lanka and traveled a distance of $s_3 = v_3(t_3-t_0) = 1,756\,\text{km}$.

(i) The increase of the height of the antinode of the tsunami when it approaches the coast is a consequence of energy conservation. The flow of energy of the wave remains constant. Equations (4.67) and (4.68) show that the energy of the wave is proportional to $(\Delta y_0)^2$, which implies that the flow of energy is proportional to $(\Delta y_0)^2 v$, which implies $\Delta y_0 D^{1/4} = const$. It follows from the last equation that $\left(\frac{\Delta y_{01}}{\Delta y_{02}}\right)^4 = \frac{D_2}{d_1}$, which leads to $D_2 = 11.6\,\mathrm{m}$. The speed of tsunami for this depth is $v = \sqrt{gD_2} = 10.7\,\mathrm{m/s}$, and it is therefore almost impossible to run away from the tsunami if you are close to the coast. On the other hand, the modest height of the wave away from the coast opens the way for more efficient defense strategies for boats (even for people) moving to the region with deeper water.

We refer interested readers to references [11] and [22]. The authors presented part of this problem at the national physics competition for the third grade of high school in Serbia in 2017.

Problem 31   Earthquake

Why do earthquakes occur?
What is the time interval between two earthquakes and what is the duration of an earthquake?
How can we model irregularities of time intervals between earthquakes?

Earthquakes appear due to the motion of tectonic plates in the Earth's lithosphere. Elastic energy is accumulated in these plates during their slow motion. In the moment when the elastic force becomes larger than the friction force between tectonic plates, sliding between the plates occurs, which is accompanied by the release of a large amount of energy. At the same time, seismic waves are created that are detected at the surface of the Earth as an earthquake.

(a) The typical relative elongation of a tectonic plate in the moment when an earthquake occurs is $\varepsilon = \frac{\Delta l}{l} = 3.00 \cdot 10^{-5}$, while the yearly change of its relative elongation is $\chi = 3.00 \cdot 10^{-7}\,\mathrm{yr}^{-1}$, where the unit yr denotes one year. Assume that the tectonic plate returns to the state without elongation after each earthquake. Calculate the time interval between two earthquakes caused by this tectonic plate.

To give a more quantitative description of the emergence of earthquakes, we introduce a simple model of a tectonic plate in which it is described by a body of mass $m = 1.00 \cdot 10^{10}\,\mathrm{kg}$ connected to a spring of stiffness $k = 1.30 \cdot 10^{10}\,\frac{\mathrm{N}}{\mathrm{m}}$, which models the elastic properties of the plate. The body is positioned on another plate that moves at a constant speed $v = 3.00\,\mathrm{cm/yr}$, as shown in Figure 4.42. The coefficient of static friction between the plates is $\mu_{st} = 0.600$, while the coefficient of dynamic friction is $\mu_{\text{dyn}} = 0.400 < \mu_{\text{st}}$. In the initial moment the spring is not elongated and the body does not slide on the plate. The system is in a gravitational field with a magnitude of $g = 9.81 \, \frac{\text{m}}{\text{s}^2}$.

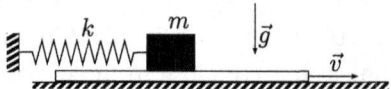

Figure 4.42 The model of tectonic plates that describes the occurence of earthquakes

(b) Show that the motion of the body of mass m becomes periodic after some time. Determine the period of these oscillations.

(c) Calculate the time interval within one period when the system is in the regime where the body of mass m slides on the plate.

(d) Calculate the amount of heat released due to sliding between the body of mass m and the plate during one period.

(e) Calculate the change in the temperature of the top tectonic plate assuming the total heat that is released leads to its heating. The specific heat capacity of the plate is $c = 800 \, \frac{\text{J}}{\text{kg·K}}$.

In the previous model earthquakes occur periodically, which is a consequence of the simplicity of the model. It is known that earthquakes occur in irregular time intervals and that it is practically impossible to predict them. To describe irregularity in the occurrence of earthquakes, we introduce the following model of a box with balls.

We consider a box that consists of N cells. Initially all cells are empty. After each time step $\Delta\tau$ we randomly choose one cell and put the ball in it. If there is already a ball in the cell, we do not put the new ball. The process is continued until all cells are filled. The box is then emptied and the whole process is repeated.

The box considered is a simplified model of a tectonic plate; adding the balls models the accumulation of elastic energy in the plate, while emptying the box corresponds to the release of energy during an earthquake.

(f) Determine the mean value and standard deviation of the time interval between the moment when the i-th ball was put in the box and the moment when the $(i+1)$-th ball was put in the box.

(g) Determine the mean value and standard deviation of the time necessary to fill all cells. Assume that $N > 10$.

(h) In the small town of Parkfield, California, earthquakes occurred on January 9, 1857, February 2, 1881, March 3, 1901, March 10, 1922, June 8, 1934, June 28, 1966, and September 28, 2004. The time intervals between these earthquakes (in years) are therefore $c_1 = 24.07$, $c_2 = 20.08$, $c_3 = 21.02$, $c_4 = 12.25$,

$c_5 = 32.05$, and $c_6 = 38.25$. Assume that the moments of earthquake occurrence can be well described using the model of the box with balls. Calculate the parameters N and $\Delta\tau$ of the model.

The square of the standard deviation is by definition equal to the mean value of the square of deviation of a physical quantity from its mean value.
You can use the following identities for $0 < q < 1$:

$$\sum_{k=1}^{\infty} kq^k = \frac{q}{(1-q)^2}, \tag{4.70}$$

$$\sum_{k=1}^{\infty} k^2 q^k = \frac{q(q+1)}{(1-q)^3}. \tag{4.71}$$

You can use the following approximations for $N > 10$:

$$\sum_{i=0}^{N-1} \frac{N}{N-i} \approx N(C + \ln N) + \frac{1}{2} \tag{4.72}$$

and

$$\sum_{i=0}^{N-1} \frac{iN}{(N-i)^2} \approx N^2 \left[\frac{\pi^2}{6} - \frac{1+C+\ln N}{N} \right], \tag{4.73}$$

where $C = 0.5772157$ is the Euler constant.

Solution of Problem 31

(a) The relative elongation of the tectonic plate in the moment when the earthquake occurs is $\varepsilon = T\chi$, where T is the period between two earthquakes. Consequently $T = \frac{\varepsilon}{\chi} = 100\,\mathrm{yr}$.

(b) The equation of motion of the body is $ma_x = F_{\mathrm{tr}} - kx$, where F_{tr} is the friction force between the body and the plate, x is the displacement of the body (with respect to the position where the spring is not elongated), and a_x is the acceleration of the body. Initially there is no sliding between the body and the plate. The acceleration of the body is therefore zero and consequently $F_{\mathrm{tr}} = kx$. Sliding between the body and the plate starts at time t_1 when the friction force becomes equal to the maximal value of the static friction force – that is, $\mu_{\mathrm{st}}mg = kx_1$, where x_1 is the displacement of the body in that moment; see Figure 4.43. The equation of motion after this moment is $ma_x = \mu_{\mathrm{dyn}}mg - kx$, with initial conditions $v_x(t_1) = v$ and $x(t_1) = x_1$. This equation describes harmonic oscillation of the body at an angular frequency $\omega = \sqrt{\frac{k}{m}}$ where equilibrium position of the body x_2 satisfies the condition $\mu_{\mathrm{dyn}}mg = kx_2$, while we find the amplitude

A from the fact that the energy in amplitude position is equal to the energy at position x_1:

$$\frac{1}{2}kA^2 = \frac{1}{2}k(x_1 - x_2)^2 + \frac{1}{2}mv^2. \qquad (4.74)$$

In the previous equation, the term on the left-hand side is the energy of the oscillator in the amplitude position. The first term on the right-hand side is the potential energy when the body is at x_1 and therefore it is at a distance $|x_1 - x_2|$ from equilibrium position, while the second term on the right-hand side is the kinetic energy in that position. Sliding of the body on the plate stops at time τ_1 when the velocity is again equal to $v_x = v$; see Figure 4.43. This happens when the body is at the same distance from equilibrium as in the moment when sliding starts. The position x' of the body in that moment is found from the condition $x' - x_2 = -(x_1 - x_2)$, which leads to $x' = 2x_2 - x_1$.

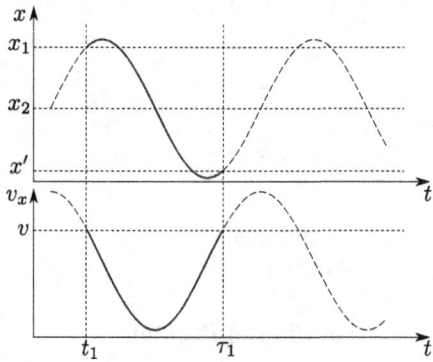

Figure 4.43 The graphs of the dependences $x(t)$ and $v_x(t)$ in the time interval from t_1 to τ_1 are represented by a full line. Dashed lines are used for easier view of the figures.

To find the time interval Δt_a when sliding regime is present, we start from the equation $x(t) = x_2 + A\sin(\omega t + \phi)$, which implies that the velocity is $v_x(t) = A\omega\cos(\omega t + \phi)$. The moments t_n and τ_n that satisfy $v_x(t) = v$ are determined from $\omega t_n + \phi = \arccos\left(\frac{v}{A\omega}\right) + 2n\pi$ and $\omega\tau_n + \phi = 2\pi - \arccos\left(\frac{v}{A\omega}\right) + 2n\pi$. At times t_n we have $x > x_2$, while at time τ_n we have $x < x_2$. Therefore, $\Delta t_a = \tau_n - t_n = \frac{2\pi}{\omega} - \frac{2}{\omega}\arccos\left(\frac{v}{A\omega}\right)$. The graphs of the dependence $x(t)$ and $v_x(t)$ are shown in Figure 4.43.

After the body reaches position x' it again moves at a constant velocity v until it comes again to the position x_1. The time interval between these two moments is $\Delta t_b = \frac{x_1 - x'}{v}$. The body then slides again until it reaches the position x' and the whole cycle repeats. The period of this motion is $T = \Delta t_a + \Delta t_b =$

$2\sqrt{\frac{m}{k}}\left(\pi - \arccos\frac{v}{\omega A}\right) + \frac{2\left(\mu_{st} - \mu_{dyn}\right)mg}{kv}$, where $\omega = 1.14\,\mathrm{s}^{-1}$, while A is obtained from equation (4.74) and is equal to $A = 1.51\,\mathrm{m}$, which leads to $T = 101\,\mathrm{yr}$.

(c) The time interval in question is $\Delta t_a = \frac{2\pi}{\omega} - \frac{2}{\omega}\arccos\left(\frac{v}{A\omega}\right) = 2\sqrt{\frac{m}{k}}\left(\pi - \arccos\frac{v}{\omega A}\right) = 2.76\,\mathrm{s}$. The result implies that the duration of the earthquake is much shorter than the time between two earthquakes, in agreement with observations from nature.

(d) The body is sliding on the plate in the time interval from $t = t_1$ to $t = \tau_1$. The work performed by the friction force on the body of mass m in that interval is $A_1 = \int_{t_1}^{\tau_1} \mu_{dyn}mg \cdot dx$, which leads to $A_1 = \mu_{dyn}mg\left[x(\tau_1) - x(t_1)\right]$, and consequently $A_1 = -2\mu_{dyn}mg\left(x_1 - x_2\right) = -2\mu_{dyn}\left(\mu_{st} - \mu_{dyn}\right)\frac{m^2g^2}{k}$. The work performed by the friction force on the bottom plate during the same time interval is $A_2 = -\mu_{dyn}mg \cdot v\Delta t_a$. The amount of heat released during that time interval is therefore

$$Q = -\left(A_1 + A_2\right) = 2\mu_{dyn}\left(\mu_{st} - \mu_{dyn}\right)\frac{m^2g^2}{k} + \mu_{dyn}mg \cdot v\Delta t_a = 1.18 \cdot 10^{11}\,\mathrm{J}.$$

$$(4.75)$$

(e) From the equation $Q = mc\Delta t$, we find that the temperature of the body will increase by $\Delta t = \frac{Q}{mc} = 1.48 \cdot 10^{-2}\,^\circ\mathrm{C}$.

(f) The probability that the $(i+1)$-th ball will be put in the cell after time interval $\Delta\tau$ is equal to the ratio of the number of empty cells and the total number of cells $p_1 = p = \frac{N-i}{N}$. The probability that this ball will be put in the cell after time interval $2\Delta\tau$ is $p_2 = (1-p)p$, the same probability after time interval $3\Delta\tau$ is $p_3 = (1-p)^2 p$, while for time interval $k\Delta\tau$ it is $p_k = (1-p)^{k-1}p$. The average time in question is then by definition $t_i = \sum_{k=1}^{\infty} kp_k\Delta\tau$. Using the identity $\sum_{k=1}^{\infty} k(1-p)^{k-1} = \frac{1}{p^2}$ we get $t_i = \frac{\Delta\tau}{p} = \frac{N\Delta\tau}{N-i}$.

The standard deviation of the time in question is determined by the expression $\sigma_i^2 = \sum_{k=1}^{\infty}\left(k - \frac{1}{p}\right)^2(\Delta\tau)^2 p_k$. We then have $\sigma_i^2 = (\Delta\tau)^2 \sum_{k=1}^{\infty}\left[k^2 p(1-p)^{k-1} + \frac{1}{p}(1-p)^{k-1} - 2k(1-p)^{k-1}\right]$. Using the identities given, it follows that $\sigma_i^2 = (\Delta\tau)^2 \frac{1-p}{p^2}$ - that is, $\sigma_i = \Delta\tau\frac{\sqrt{1-p}}{p}$.

(g) The average time it takes to fill all cells is $T = \sum_{i=0}^{N-1} t_i = \sum_{i=0}^{N-1}\frac{N}{N-i}\Delta\tau$ - that is,

$$T = \Delta\tau\left[N(C + \ln N) + \frac{1}{2}\right].$$

$$(4.76)$$

The standard deviation of this time is by the definition given as $\sigma^2 = \left\langle\left(\sum_{i=0}^{N-1} t_i - \left\langle\sum_{i=0}^{N-1} t_i\right\rangle\right)^2\right\rangle$, where $\langle\ldots\rangle$ denotes the average. The last expression can be transformed to $\sigma^2 = \left\langle\left[\sum_{i=0}^{N-1}\left(t_i - \langle t_i\rangle\right)\right]^2\right\rangle$. We then obtain

$\sigma^2 = \left\langle \sum_{i=0}^{N-1} \left(t_i - \langle t_i \rangle \right)^2 + \sum_{i \neq j} \left(t_i - \langle t_i \rangle \right) \left(t_j - \langle t_j \rangle \right) \right\rangle$. Since the times t_i and t_j are mutually independent, it follows that $\langle t_i t_j \rangle = \langle t_i \rangle \cdot \langle t_j \rangle$, which leads to $\langle \left(t_i - \langle t_i \rangle \right) \left(t_j - \langle t_j \rangle \right) \rangle = 0$. The expression for standard deviation therefore reduces to $\sigma^2 = \sum_{i=0}^{N-1} \sigma_i^2 = \sum_{i=1}^{N-1} \frac{iN}{(N-i)^2} (\Delta \tau)^2 = N^2 (\Delta \tau)^2 \left[\frac{\pi^2}{6} - \frac{1+C+\ln N}{N} \right]$, and finally

$$\sigma = \Delta \tau N \sqrt{\frac{\pi^2}{6} - \frac{1+C+\ln N}{N}}. \tag{4.77}$$

(h) We find from the given data that the average time interval between two earthquakes is $T = \frac{1}{6} \sum_{i=1}^{6} c_i = 24.62\,\mathrm{yr}$, while the standard deviation is found from $\sigma^2 = \frac{1}{6} \sum_{i=1}^{6} (c_i - T)^2$, which leads to $\sigma = 8.44\,\mathrm{yr}$. Next, by dividing equations (4.76) and (4.77) it follows that

$$\frac{T}{\sigma} = \frac{N(C+\ln N) + \frac{1}{2}}{N \sqrt{\frac{\pi^2}{6} - \frac{1+C+\ln N}{N}}}, \tag{4.78}$$

– that is, $2.92 = f(N)$, where $f(N)$ is the function defined by the expression on the right-hand side of equation (4.78). The graph of that function is shown in Figure 4.44. We read from the graph that the equation $2.92 = f(N)$ has two solutions $N \approx 1.2$ and $N \approx 17$. Since it is known that $N > 10$, it follows that $N = 17$. By putting this value of N in equation (4.76) or (4.77) it follows that $\Delta \tau = 0.42\,\mathrm{yr}$.

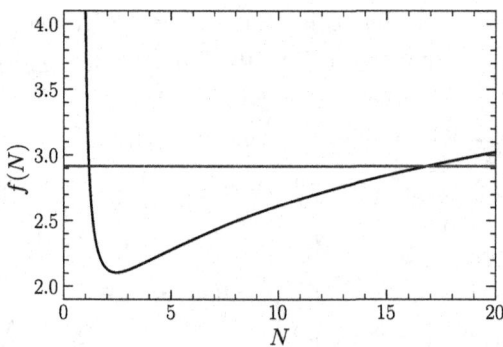

Figure 4.44 The graph of the function $f(N)$

We refer the reader interested in more details to references [12], [18], and [29]. The authors presented part of this problem at the competition for the selection of the Serbian team for the European Physics Olympiad in 2020.

5

Energy

Problem 32   War of the Currents

Who Won the War of the Currents and Why: Tesla or Edison, AC or DC?

Electrical energy is produced in generators in power stations. It is then transmitted via overhead power lines to customers. Power stations are usually located far away from customers. For this reason, the loss of electrical energy in the power lines can be large. This fact caused the so called "war of the currents" between Nikola Tesla and Thomas Edison in the late 1880s. Nikola Tesla advocated his efficient system that used alternating current, while Thomas Edison introduced his transmission system based on direct current. We consider both transmission systems and use as an example one American city from that period that used electrical energy for lighting only. Incandescent bulbs were used for lighting that required a voltage of $U = 100\,\text{V}$. The power required by the city was $P = 10.0\,\text{kW}$. It was transmitted using two-wire power lines of length $l = 2.00\,\text{km}$, made of copper wires whose cross section is $S = 0.800\,\text{cm}^2$ and specific resistance is $\rho = 1.78 \cdot 10^{-8}\,\Omega\text{m}$.

Since there was no efficient way to change the voltage in a DC circuit, the voltage on transmission lines in Edison's system was the same as the voltage used by the customers.

(a) Calculate the ratio of the power lost in the power lines during transmission of DC current from the power station to the city and the power required by the city.

(b) Calculate the power the power station needs to generate in order to satisfy the needs of the city.

In Tesla's transmission system the voltage at the output of the generators in power station U_1 is increased several thousand times using a transformer. The voltage at the secondary of this transformer is U_2. The transmission lines are then at high voltage. This voltage is then decreased to the required value U using a second transformer at the customer's side. Assume that the transformers are ideal.

(c) Calculate the power the power station to needs to generate in order satisfy the needs of the city if Tesla's transmission system is used. The effective value of the voltage at the secondary of the first transformer is $U_2 = 10\,\text{kV}$.

(d) How many times smaller are power losses in transmission lines when Tesla's system is used in comparison to Edison's system?

(e) Calculate the difference of voltages at the secondary of the first transformer and the primary of the second transformer.

(f) Calculate the number of turns of the secondary of the second transformer if its primary has $N_2 = 5,000$ turns.

During the flow of current through the transmission line the insulation of the wire melted, which effectively led to the appearance of finite resistance between the two lines at that point, as shown in Figure 5.1. The failure was reported and the electricians decided to use the following procedure to determine the position of the failure and to repair the system. They disconnected the cables from the transformers and then they performed the following three measurements. First, they measured the resistance between points T_2' and T_2'', when T_1' and T_1'' are not connected, and they obtained $R_1 = 0.68\,\Omega$. Next they measured the resistance between T_2' and T_2'' when T_1' and T_1'' are short-circuited and obtained $R_2 = 0.38\,\Omega$. Finally they measured the resistance between T_1' and T_1'' when T_2' and T_2'' are disconnected and got $R_3 = 0.66\,\Omega$.

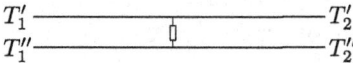

Figure 5.1 Two-wire cable that connects the first and the second transformers

(g) What is the distance between the place of failure and the second transformer?

Solution of Problem 32

(a) The DC current I that flows through the transmission line is $I = P/U$, while the resistance of the transmission line is $R = 2\frac{\rho l}{S} = 0.890\,\Omega$. Factor 2 originates from the fact that the line consists of two wires. The power lost in the transmission is:

$$P_g = I^2 R = \frac{2P^2\rho l}{U^2 S}, \tag{5.1}$$

which leads to

$$\frac{P_g}{P} = \frac{2P\rho l}{U^2 S} = 0.890. \tag{5.2}$$

(b) The power station should produce power $P_e = P + P_g = P + 0.89P = 1.89P = 18.9\,\text{kW}$.

(c) Since the transformers are ideal, the power produced by the power station is $P'_e = U_2 I_2$ in this case, where I_2 is the current through the transmission lines. The power lost in the transmission lines is $P'_g = I_2^2 R$, which leads to

$$P'_e = P + P'_g = P + P'^2_e \frac{R}{U_2^2} \approx P = 10.0\,\text{kW}. \tag{5.3}$$

The second term in the equation (5.3) was neglected since $\frac{R}{U_2^2} \to 0 \frac{1}{\text{W}}$. The result implies that the losses in transmission lines in Tesla's system are practically negligible.

(d) It follows from the solution of part (c) that the losses in the transmission lines in Tesla's system are $P'_g = P^2 \frac{R}{U_2^2}$, which leads to

$$\frac{P_g}{P'_g} = \frac{U_2^2}{U^2} = 10{,}000. \tag{5.4}$$

(e) The voltage U_2 on the secondary of the first transformer and the voltage U'_2 on the primary of the second transformer differ by

$$\Delta U_2 = I_2 R = \frac{P}{U_2} \frac{2\rho l}{S} = 0.890\,\text{V}. \tag{5.5}$$

(f) Equation (5.5) implies that the difference between U'_2 and U_2 is negligible, which leads to $N = \frac{U}{U_2} N_2 = 50$.

(g) We denote as R' the resistance caused by the failure and x the distance between the point of failure and the second transformer; see Figure 5.2. We then have $R_1 = \frac{2\rho x}{S} + R'$ for the first measurement, $R_2 = \frac{2\rho x}{S} + \frac{2\rho(l-x)R'}{SR'+2\rho(l-x)}$ for the second, and $R_3 = \frac{2\rho(l-x)}{S} + R'$ for the third. Previous equations imply that $x = \frac{R_1 - \sqrt{R_3(R_1-R_2)}}{R_1 + R_3 - 2\sqrt{R_3(R_1-R_2)}} l = 1.04\,\text{km}$.

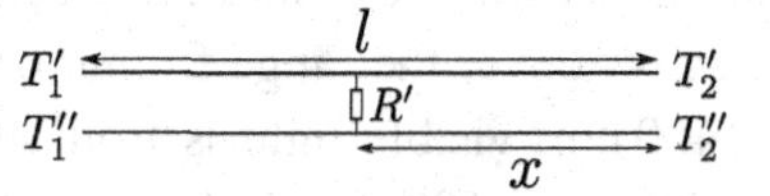

Figure 5.2 Two-wire cable that connects the first and the second transformers

The authors presented parts of this problem at a municipality-level physics competition for the third grade of high school in 2016 and at a regional physics competition for the third grade of high school in 2017 in Serbia.

Problem 33   Greenhouse Effect

How does the Earth's atmosphere affect the temperature of the Earth's surface?

Thermal radiation of the sun is the main source of energy on the Earth. Therefore, the temperature on the surface of the Earth is mainly determined by energy emitted from the sun that reaches the Earth's surface. On the other hand, processes of absorption and emission of electromagnetic radiation in the Earth's atmosphere also affect the temperature on the surface of the Earth. For this reason, changes in chemical composition of the atmosphere caused by emission of certain gases, such as CO_2, may increase the temperature of the Earth's surface. We investigate the effects on the temperature of the Earth's surface in this problem.

Assume that the Earth orbits around the sun on a circular trajectory of radius $R = 1.495 \cdot 10^{11}$ m and that the sun emits as a black body of temperature $T_s = 5,780$ K. The radius of the sun is $R_s = 6.96 \cdot 10^8$ m. The Stefan–Boltzmann constant is $\sigma = 5.670 \cdot 10^{-8} \frac{W}{m^2 K^4}$, and Wien's constant is $b = 2.898 \cdot 10^{-3}$ m $\cdot$ K.

(a) Calculate the mean power per unit of the Earth's surface of the sun's radiation that is incident on the Earth's surface. Assume in this part of the problem that the Earth does not have an atmosphere.

In the rest of this problem assume that the temperature of the atmosphere does not depend on altitude, that the only way of heat exchange is electromagnetic radiation, and that the Earth's surface emits radiation as a black body.

(b) Assume that $\alpha = 30\%$ of the sun's radiation that reaches the Earth's atmosphere is reflected at clouds in lower parts of the atmosphere. Assume also that all radiation emitted from the Earth's surface passes through the atmosphere without absorption. Calculate the temperature of the Earth's surface.

(c) Calculate the wavelength at which spectral radiance of the sun's (the Earth's) radiation has a maximum. In what region of the electromagnetic spectrum are these wavelengths? X-rays cover the range from 0.01 to 1 nm, the ultraviolet part is from 1 nm to 400 nm, visible light is from 400 nm to 700 nm, infrared radiation is from 700 nm to 100 μm, and microwaves are from 100 μm to 100 mm.

The Earth's atmosphere has a complex chemical composition and absorption of electromagnetic radiation strongly depends on the wavelength. In the rest of the problem, we use a simplified model for interaction of electromagnetic radiation with the Earth's atmosphere. We assume that the atmosphere is a gray body transparent for visible light, while it absorbs $\eta = 75\%$ of infrared radiation. Assume again that $\alpha = 30\%$ of the sun's radiation is reflected on clouds in the lower parts of the atmosphere.

(d) Calculate the temperature of the Earth's surface within this model of the atmosphere. What is the difference of this temperature and the temperature determined in part (b)? The increase of the temperature of the Earth's surface due to absorption of the Earth's electromagnetic radiation in the atmosphere is called the greenhouse effect.

(e) Assume that η increases from $\eta = 75\%$ to $\eta = 80\%$ due to the larger presence of CO_2 in the atmosphere. Calculate the change in the temperature of the Earth's surface due to this effect.

Solution of Problem 33

(a) According to the Stefan–Boltzmann law, the power of the sun's radiation is $P = \sigma T_s^4 \cdot 4R_s^2 \pi$. The sun's radiation is isotropic. Therefore, the fraction of the sun's radiation that reaches the Earth's surface is equal to the ratio of the solid angle subtended by the Earth to an observer from the sun and the full solid angle $u = \frac{R_z^2 \pi / R^2}{4\pi}$, where R_z is the radius of the Earth. The mean power per unit of the Earth's surface in question is $S = \frac{uP}{4R_z^2 \pi}$ and consequently $S = \frac{\sigma T_s^4 R_s^2}{4R^2} = 342.9\,\frac{W}{m^2}$.

(b) The fraction $1 - \alpha$ of the sun's radiation that reaches the upper layers of the atmosphere is absorbed on the Earth's surface. In steady state, the power of this radiation is equal to the power of radiation emitted from the Earth's surface. Consequently $S(1 - \alpha) = \sigma T_z^4$, which leads to

$$T_z = \sqrt[4]{\frac{S(1 - \alpha)}{\sigma}} = 255.1 \text{ K}. \qquad (5.6)$$

(c) The wavelengths at which the spectral radiance of the sun's (the Earth's) radiation has a maximum is given by Wien's law. This wavelength is equal to $\lambda_s = b/T_s = 501$ nm in the case of the sun and $\lambda_z = b/T_z = 11.4\,\mu m$ in the case of the Earth. Consequently the sun predominantly radiates in the visible part of the spectrum, while the Earth radiates in infrared.

(d) The Earth's surface absorbs the power per unit surface $S(1 - \alpha)$ that originates from the sun's radiation, as well as $\eta \sigma T_a^4$ that originates from emission of radiation by the atmosphere, where T_a is the temperature of the atmosphere.

On the other hand, the Earth emits the power per unit surface of σT_z^4. Total emitted power must be equal to total absorbed power in steady state, hence

$$S(1-\alpha) + \eta \sigma T_a^4 = \sigma T_z^4. \tag{5.7}$$

Next, at the surface between the upper layer of the atmosphere and outer space, we have the following balance. The power per unit surface S comes from outer space, the reflected power αS leaves the Earth, the power radiated by the atmosphere $\sigma T_a^4 \eta$ also leaves the Earth, as well as the power $\sigma T_z^4 (1-\eta)$ radiated by the Earth's surface reduced by the part absorbed in the atmosphere. Since the power that leaves the Earth at this surface must be equal to the power that comes from outer space, we obtain

$$S = \alpha S + \sigma T_a^4 \eta + \sigma T_z^4 (1-\eta). \tag{5.8}$$

From equations (5.7) and (5.8) we find

$$T_z = \sqrt[4]{\frac{S(1-\alpha)}{\sigma \left(1 - \frac{\eta}{2}\right)}} = 286.9 \, \text{K}. \tag{5.9}$$

This temperature is $31.8 \, \text{K}$ larger than the temperature determined in part (b).

(e) Using equation (5.9) we find that in this case $T_z = 289.8 \, \text{K}$. Therefore, the temperature increases by $2.9 \, \text{K}$.

We refer the reader interested in the thermal properties of the Earth's atmosphere to reference [2]. This problem was created by modifying and extending the problem Mihailo Rabasović presented at a regional physics competition for the fourth grade of high school in Serbia in 2012.

Problem 34   Wind Turbines

What is the theoretical limit for conversion of wind energy to electrical energy?

How much energy does a wind farm produce?

The European Union plans to increase the contribution of renewable energy sources to total energy production to 25% by 2025, while it is expected that it should reach almost 100% by the end of this century. Wind energy is one of the most economical types of renewable energy. Wind turbines are used to exploit the energy of the wind. The blades of a wind turbine are designed to rotate when the wind is blowing. Therefore, the energy of the wind is converted into mechanical energy of the blades. Next generators convert this energy to electrical energy. The goal is to maximize the efficiency of the wind turbine (defined as the ratio of the mechanical energy of the blades and the energy the wind used to move the blades). This efficiency is not equal to 100% even in an ideal case. The theoretical maximum efficiency of an ideal wind turbine is called the Betz limit in honor of German physicist Albert Betz, who calculated this value in 1982.

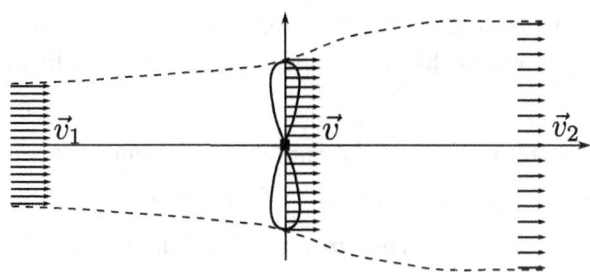

Figure 5.3 Scheme of wind flow around an ideal wind turbine

The wind is flowing at a speed v_1 at atmospheric pressure in a region far away from the turbine. This speed reduces near the turbine and the air passes through the turbine at speed v. An ideal wind turbine consists of an infinite number of blades and can be considered as a thin disk with holes that is perpendicular to the x-axis (Figure 5.3). The speed of air further reduces after it passes through the turbine to speed v_2 far away from the turbine where the pressure is atmospheric. Assume that air is an ideal incompressible fluid of density $\rho = 1.22 \, \text{kg/m}^3$ and constant temperature.

(a) Express the speed v of air when it passes through the turbine in terms of speeds v_1 and v_2.
(b) Calculate the efficiency of an ideal wind turbine in terms of the parameter $b = v_2/v_1$.
(c) Determine the Betz limit. What is the value of the parameter b_{Betz} in that case?
(d) The Betz limit resembles somewhat the efficiency of the Carnot cycle in thermodynamics, which describes the maximal work that can be produced by a fluid in a closed thermodynamic cycle. What is the efficiency of the Carnot cycle with the ratio of minimal and maximal temperature equal to b_{Betz}? What is more efficient, an ideal wind turbine or an ideal heat engine?

The efficiency of commercial wind turbines is smaller than the Betz limit. Nevertheless, the amount of electrical energy they can produce is quite significant. In reality, when the wind reaches the blade the flow of air becomes turbulent behind the blade in the region up to one half of its length in the axial direction. When the blades rotate fast, the wind effectively reaches a big flat disk that creates a big resistance force of the turbulent air. Consequently the blades can be damaged due to the dynamic load. On the other hand, if the blades are rotating slowly, the wind can pass between them without any effect and only a small part of energy of the wind is used (well below the Betz limit). Therefore, the ratio of the speed of the top of the blade and the speed of the wind (lambda coefficent) is of significant importance for the efficiency of commercial wind turbines. The efficiency of the wind turbine is maximal when the lambda coefficient is optimal – that is, when local turbulence due to contact of air and the blade disappears before the next blade arrives to this region.

(e) Calculate the optimal value of the lambda coefficient for a wind turbine with $n = 3$ blades.

The dependence of the efficiency of a wind turbine on the lambda coefficient is shown in Figure 5.4, for a wind turbine with $n = 3$ blades. It can be concluded from the figure that there is an optimal speed of rotation of the blades for each wind speed. Modern wind turbines are designed so that they can have different speeds of rotation of the blades. However, traditional wind turbines are designed so that the blades rotate at a constant angular velocity, which leads to a constant frequency of the alternating current that is produced.

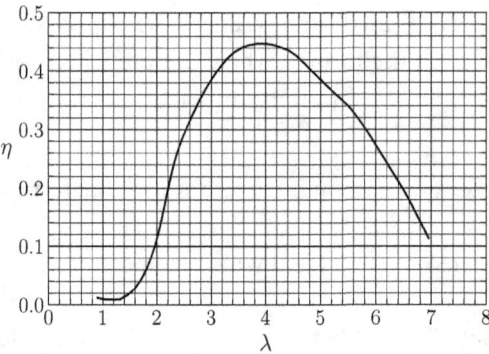

Figure 5.4 The characteristics of a wind turbine

Consider a wind farm that contains $N = 30$ wind turbines the characteristics of which are shown in Figure 5.4. The length of the blades of the wind turbine is $l = 3.5$ m and they rotate at constant angular velocity of magnitude $\omega = 4$ rad/s. Wind resources at the location of this wind farm are shown in Figure 5.5 where the histogram of annual frequency of appearance of different wind speeds is shown.

(f) Plot the histogram of energy produced by one wind turbine for different wind speeds.

(g) Calculate the energy produced in this wind farm in a year.

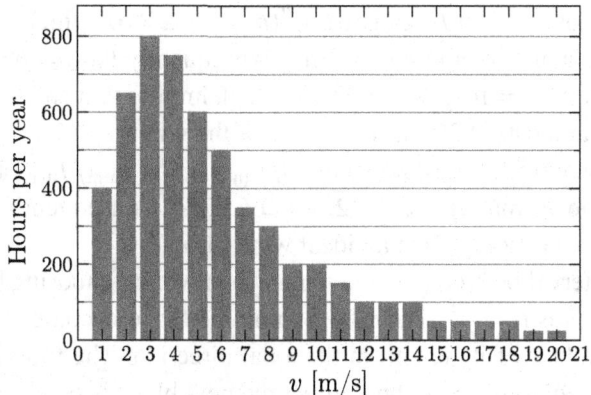

Figure 5.5 Histogram of annual frequency of appearance of difference wind speeds

Solution of Problem 34

(a) When a small part of air of mass Δm passes through an ideal wind turbine, its energy changes by

$$\Delta E = \frac{1}{2}\Delta m(v_2^2 - v_1^2) = -Fv\Delta t, \tag{5.10}$$

where F is the force of the air on the wind turbine and Δt is the time it takes for this part of the air to pass through the wind turbine at a speed v. In a similar manner, the change of the momentum of this part of the air is

$$\Delta p = \Delta m(v_2 - v_1) = -F\Delta t. \tag{5.11}$$

Dividing equations (5.10) and (5.11) we get

$$v = \frac{1}{2}(v_1 + v_2). \tag{5.12}$$

(b) The power invested is the power of the wind when there is no wind turbine – that is, $P_u = \frac{1}{2}\rho S v_1^3$, where S is the area of the cross section of the wind turbine. Useful power is the power of the wind turbine $P_k = Fv$. Using equations (5.10) and (5.12) we get $P_k = \frac{1}{4}\rho S(v_1^2 - v_2^2)(v_1 + v_2)$. The efficiency of an ideal wind turbine is

$$\eta(b) = \frac{P_k}{P_u} = \frac{1}{2}(1 + b)(1 - b^2). \tag{5.13}$$

(c) The Betz limit is the maximal value of the function $\eta(b)$. The function $\eta(b)$ reaches maximum when $\eta'(b) = \frac{1}{2}(-3b^2 - 2b + 1) = 0$, which leads to two solutions: $b_1 = 1/3$ and $b_2 = -1$. The second solution is discarded since b is a positive number equal to the ratio of the speeds. The second derivative of the function $\eta(b)$ at b_1 satisfies $\eta''(b_1) = -2 < 0$, which implies that the function has a maximum at this point. Consequently $b_{Betz} = b_1 = 1/3$ and the Betz limit is $\eta_{Betz} = \eta(b_{Betz}) = 0.593$, which implies that an ideal wind turbine can exploit at most 59.3% of the energy of the wind.

(d) The efficiency of the Carnot cycle is $\eta_{Carnot} = 1 - T_{min}/T_{max}$. If $T_{min}/T_{max} = b_{Betz}$, it follows that $\eta_{Carnot} = 2/3 = 66.67\%$. Consequently an ideal heat engine is more efficient than an ideal wind turbine.

(e) The time interval between the moments when two neighboring blades pass the same position is $t_1 = \frac{2\pi}{n\omega}$. Air flow behind the blade becomes turbulent in the region up to half of its length in the axial direction. The time it takes for the turbulence to dissappear and not affect the next blade is $t_2 = \frac{l}{2v}$, where l is the length of the blade. The optimal value of the lambda coefficient is obtained when these times are the same, which implies $\lambda = \frac{l\omega}{v} = \frac{4\pi}{n} = 4.19$.

(f) To plot the histogram in question, the energy produced for each speed should be calculated. The results are shown in the following table:

$v[m/s]$	$\Delta t[h]$	λ	η	$P_v[kW]$	$E[kWh]$
1	400	14.0	0	0.023	0
2	650	7.00	0.1	0.189	12.21
3	800	4.67	0.41	0.634	207.9
4	750	3.50	0.43	1.502	484.5
5	600	2.80	0.35	2.934	616.2
6	500	2.33	0.22	5.071	557.7
7	350	2.00	0.12	8.052	338.2
8	300	1.75	0.04	12.02	144.2
9	200	1.56	0.01	17.11	34.23
10	200	1.40	0.01	23.47	46.95
11	150	1.27	0.01	31.25	46.87
12	100	1.17	0.01	40.57	40.57
13	100	1.08	0.01	51.58	51.58
14	100	1.00	0.01	64.42	64.42
15	50	0.93	0.00	79.23	0
16	50	0.87	0.00	96.16	0
17	50	0.82	0.00	115.3	0
18	50	0.78	0.00	136.9	0
19	25	0.74	0.00	161.0	0
20	25	0.70	0.00	187.8	0

For each speed v_i, $i = \{1, 20\}$ we read Δt_i from the histogram in Figure 5.5 and we calculate $\lambda_i = l\omega/v_i$. Then we read the value of η_i from Figure 5.4. Next we calculate the power of wind $P_{vi} = \frac{1}{2}\rho S v_i^3 = \frac{\pi}{2}\rho l^2 v_i^3$ and the energy produced $E_i = \eta P_{vi}\Delta t_i$. The dependence $\eta(\lambda)$ suggests that the wind turbine is designed not to operate at speeds smaller than $2\,\mathrm{m/s}$ and larger than $14\,\mathrm{m/s}$. Yearly production of energy can be represented using a histogram in Figure 5.6.

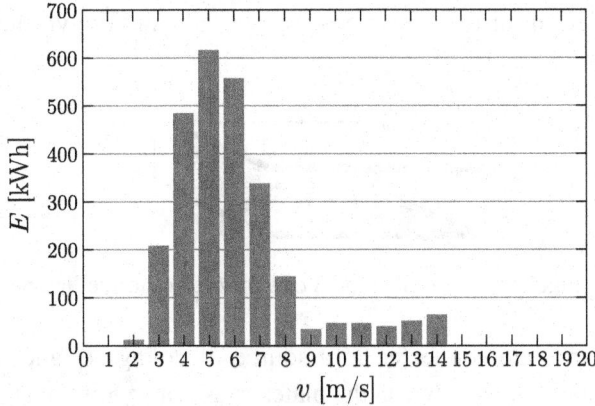

Figure 5.6 Histogram of yearly production of energy of a wind turbine

(g) Total yearly production of energy of a wind turbine E_{1uk} is obtained when we add the energies produced for each speed of the wind $E_{1uk} = \sum_{i=1}^{20} E_i = 2.65\,\mathrm{MWh}$. Total energy produced in the wind farm is $E_{uk} = N E_{1uk} = 79.4\,\mathrm{MWh}$.

Problem 35   Piezoelectric Oscillator

How does a quartz watch convert electrical energy to mechanical energy?
How does the accuracy of a quartz watch depend on the temperature of the environment?

The piezoelectric effect is the occurrence of voltage due to mechanical stress in some types of crystals. The effect is reversible, which means that mechanical deformation of the crystal occurs due to applied voltage. Jacques and Pierre Curie discovered the effect in 1880. The most well-known piezoelectric materials are quartz, topaz, Rochelle salt, and some organic macro-molecules (dentine, DNA). The devices that convert mechanical energy to electrical and vice versa are widely used nowadays. One of the most important devices of this type is the quartz oscillator. It exploits the mechanical resonance of a vibrating piezoelectric crystal to create an electrical signal with a very precise frequency. This effect is used in quartz watches, digital integrated circuits in computers, mobile phones, and radio transmitters and receivers.

We first consider a simplified model of a piezoelectric oscillator. Two metallic plates, both of area S and mass m, are positioned in horizontal planes in the gravitational field (Figure 5.7). The crystal positioned between the plates of the piezoelectric oscillator can be modeled as n springs made of insulating material. The stiffness of each spring is k. The bottom plate is fixed at a horizontal surface, while the equilibrium distance between the plates is X_0. The vacuum permittivity is ε_0.

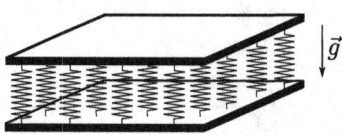

Figure 5.7   Simple model of a piezoelectric oscillator

(a) The plates are then connected to a source of voltage U and form a capacitor. Electrostatic force between the plates causes an additional displacement of the upper plate and changes the equilibrium distance between the plates to X_1. Derive the expressions for attractive electrical force F_q and voltage U between the plates in terms of X_0, X_1, S, k, and n.

(b) The system then starts to perform harmonic oscillations when it is displaced by a small displacement x with respect to the equilibrium position at a constant voltage U. Derive the expression for the acceleration of the upper plate $\ddot{x}$ in terms of X_0, X_1, k, n, m, and x, as well as the expression for the angular frequency of small vertical oscillations ω_0 in terms of X_0, X_1, k, n, and m.

(c) An inductor with inductance L is then connected in series to a voltage source and the capacitor. Initially the capacitor is charged and the distance between its plates is X_1. Its upper plate is then displaced from the equilibrium position by a small displacement x. This system has two degrees of freedom and can be described using two parameters – the displacement of the upper plate x and the change of the capacitor charge q. Derive the expression for $\ddot{x}$ and $\ddot{q}$ in terms of X_1, S, k, n, m, L, U, x, and q.

(d) Since the oscillations of the system are harmonic, one can assume that $\ddot{x} = -\omega^2 x$ and $\ddot{q} = -\omega^2 q$. Derive the expression for angular frequency of oscillations of the system ω in terms of X_0, X_1, S, k, n, m, and L. What is the relation that the quantities X_0 and X_1 should satisfy so that the oscillations are possible?

We consider next a quartz fork, which is an example of a real piezoelectric oscillator. Quartz forks are often used as a frequency standard in watches. Their mechanical model is a harmonic oscillator in the resistive medium with a small damping coefficient $b = 7.830 \cdot 10^{-6}$ kg/s. A quartz fork is in the form of a tuning fork, but it is much smaller and it operates at ultrasound frequencies (Figure 5.8). It can be considered as an electrical device with two inputs. Each input contains a thin film that has the role of an electrode and it is used to connect the fork to the rest of the system. Mechanical oscillations of the fork create charge at electrodes. A reverse effect is obtained if voltage is applied to the electrodes. Depending on its polarity, this voltage causes either attraction or repulsion of the parts of the fork that contain the electrodes.

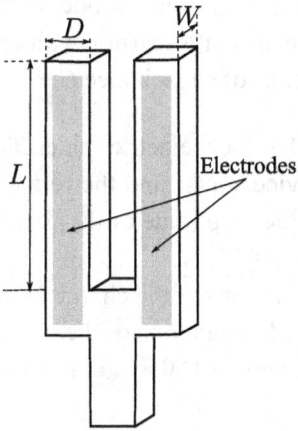

Figure 5.8 Quartz fork

When oscillation amplitudes are small, the displacements of all parts of the fork are proportional to each other. Three-dimensional motion of the fork can then be described using one coordinate only. We choose the coordinate x that describes the horizontal position of one top of the fork (the position of the other top is then $-x$). The equation that describes the oscillations of the quartz fork is then

$$\ddot{x} + \gamma \dot{x} + \Omega_0^2 x = f \cos(\Omega t + \varphi).$$

The driving force exists due to applied external voltage between the electrodes. The effective mass of the quartz fork is $m = 0.243 \rho_q V$, where $\rho_q = 2{,}659$ kg/m^3 is the density of quartz and V is the volume of one tine of the fork. The effective stiffness is $k = \frac{EWD^3}{4L^3}$, where Young's modulus of the quartz is $E = 7.87 \cdot 10^{10}$ N/m^2, while the dimensions of the tine of the fork are $W = 0.127$ mm, $D = 0.325$ mm, and $L = 2.809$ mm.

(e) Write down the equation of forced damped oscillations of the quartz fork in terms of m, k, x, b, and the driving force $F = F_0 \cos(\Omega t + \varphi)$. Calculate the values of the coefficients γ and Ω_0.

(f) Forced oscillations of the quartz fork are described by the expression $x = x_0 \cos(\Omega t + \phi)$. Find the expressions for the amplitude x_0 and the difference of phases $\delta\varphi = \phi - \varphi$ in terms of f_0, γ, Ω, and Ω_0.

The charge q at the electrodes appears due to the piezoelectric properties of quartz. This charge is proportional to the displacement of the top of the tine x with proportionality constant of $\eta = 8.130 \cdot 10^{-6} \, Cm^{-1}$. This constant depends on the geometry of the fork, the direction of cutting of the crystal, and the shape and the position of the electrodes. Using the relation between the charge q and the displacement x, one can show that the mechanical model of the quartz fork is equivalent to an electric circuit that consists of a resistor with resistance R_e, the inductor of inductance L_e, and the capacitor of capacitance C_e.

(g) Making use of the fact that piezoelectric materials convert mechanical power to electrical power and vice versa, find the relation between the amplitude of the applied voltage and the amplitude of the force that acts on the tines of the fork.

(h) Find the equation of oscillations of the charge at the electrodes. Determine the expressions for R_e, L_e, and C_e in terms of the mechanical characteristics of the quartz fork and the constant η. Calculate also the numerical values of these parameters.

(i) Since the mechanical characteristics of the quartz fork can be modeled using an equivalent electric circuit, it turns out that the complete model of the quartz fork can be replaced by an equivalent electric circuit. Draw the scheme of this electrical circuit.

It was experimentally determined that the frequency of the fork depends on temperature. This dependence is described by a parabolic law. The graph of this dependence is shown in Figure 5.9. A quartz fork with frequency $f = 2^{15} \, Hz$ was used in the experiment.

(j) At which temperature is the time shown by a quartz watch with such a fork accurate? Will this watch be late or early if it is used at much smaller and much larger temperatures?

(k) Determine the empirical formula for the dependence of the change of frequency of the quartz fork on temperature.

(l) The average temperature of the room where Vladimir works is $25\,°C$, while the average temperature of the room where Nenad works is $10\,°C$. What will be the

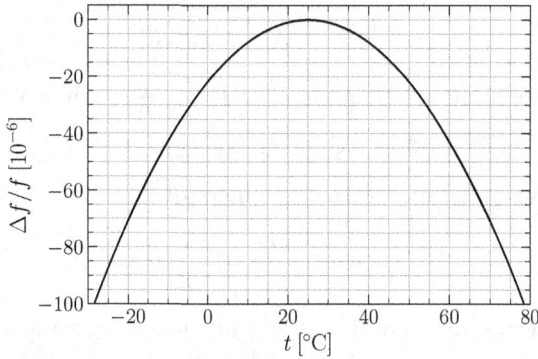

Figure 5.9 The graph of the dependence of the relative change of frequency on temperature

difference in times shown by two identical quartz clocks from Vladimir's and Nenad's offices after one year of usage?

Solution of Problem 35

(a) The gravity force acting on the top plate is in equilibrium with the elastic force of the springs, which implies $mg = nk(X' - X_0)$, where X' is the length of the spring when it is not deformed. When external voltage is applied, electrostatic force acts on the plates and the condition of equilibrium reads $mg + F_q = nk(X' - X_1)$, which leads to $F_q = nk(X_0 - X_1)$. The charge at one plate of the capacitor is $q_0 = \frac{\varepsilon_0 S}{X_1} U$. We denote as E_p the magnitude of the electric field that originates from the other plate. Gauss's law then gives $2SE_p = \frac{q_0}{\varepsilon_0}$. Electrostatic force between the plates is then $F_q = q_0 E_p = \frac{q_0^2}{2\varepsilon_0 S} = \frac{\varepsilon_0 S U^2}{2X_1^2} = nk(X_0 - X_1)$, which leads to $U = X_1 \sqrt{\frac{2nk(X_0 - X_1)}{\varepsilon_0 S}}$.

(b) When the top plate is at a distance $X_1 + x$ from the bottom one, the electrostatic force acting on it is $F'_q = \frac{\varepsilon_0 S U^2}{2(X_1 + x)^2} \approx \frac{\varepsilon_0 S U^2}{2X_1^2} - \frac{\varepsilon_0 S U^2}{X_1^3} x = F_q - \frac{\varepsilon_0 S U^2}{X_1^3} x$, while the elastic force is $F'_e = nk(X' - X_1 - x)$. Newton's second law yields $m\ddot{x} = F'_e - mg - F'_q \approx -(nk - \frac{\varepsilon_0 S U^2}{X_1^3})x = -nk(3 - 2\frac{X_0}{X_1})x$. Therefore, the angular frequency of small oscillations of the top plate is $\omega_0 = \sqrt{\frac{nk}{m}(3 - 2\frac{X_0}{X_1})}$. Small oscillations will be possible if $3X_1 > 2X_0$.

(c) Kirchhoff's second law yields $U = U_L + U_C = L\ddot{q} + \frac{(q_0 + q)(X_1 + x)}{\varepsilon_0 S}$, where $q_0 = \frac{\varepsilon_0 S U}{X_1}$ is the charge at the plates of the capacitor at initial moment of time. By neglecting the term of second-order qx, we obtain

$$\ddot{q} = -\frac{U}{LX_1}x - \frac{X_1}{\varepsilon_0 LS}q. \tag{5.14}$$

Electrostatic force at initial moment is $F_q = \frac{q_0^2}{2\varepsilon_0 S}$, while it is equal to $F_q' = \frac{(q_0+q)^2}{2\varepsilon_0 S} \approx \frac{q_0^2}{2\varepsilon_0 S} + \frac{q_0}{\varepsilon_0 S}q$ after a small displacement. The elastic force is $F_e' = nk(X' - X_1 - x)$. Newton's second law then gives

$$\ddot{x} = -\frac{nk}{m}x - \frac{U}{mX_1}q. \tag{5.15}$$

(d) By replacing $\ddot{x} = -\omega^2 x$ and $\ddot{q} = -\omega^2 q$ in equations (5.14) and (5.15), we get the following system of equations: $(\omega^2 - \frac{X_1}{\varepsilon_0 LS})q - \frac{U}{LX_1}x = 0$ and $(\omega^2 - \frac{nk}{m})x - \frac{U}{mX_1}q = 0$. This system has a nontrivial solution if $(\omega^2 - \frac{X_1}{\varepsilon_0 LS})(\omega^2 - \frac{nk}{m}) = \frac{U^2}{mLX_1^2} = \frac{2nk(X_0-X_1)}{\varepsilon_0 mLS}$. After introducing the replacement $\omega_1^2 = \frac{nk}{m}$ and $\omega_2^2 = \frac{X_1}{\varepsilon_0 LS}$, previous equation simplifies to $\omega^4 - \omega^2(\omega_1^2 + \omega_2^2) + \omega_1^2\omega_2^2(3 - 2X_0/X_1) = 0$. Physical solutions of this equation are

$$\omega_\pm = \frac{1}{\sqrt{2}}\left[\omega_1^2 + \omega_2^2 \pm \sqrt{\omega_1^4 + \omega_2^4 - 2\omega_1^2\omega_2^2(5 - 4X_0/X_1)}\right]^{1/2},$$

which implies that the system has two possible angular frequencies when the condition $\omega_\pm > 0$ – that is, $3X_1 > 2X_0$ is satisfied.

(e) The equation of forced damped oscillations of the quartz fork is $m\ddot{x} + b\dot{x} + kx = F_0\cos(\Omega t + \varphi)$. After dividing this equation by m, we obtain an equation equivalent to the equation given in the text of the problem. By comparing the corresponding terms, we obtain $\gamma = \frac{b}{m} = \frac{b}{0.243\rho_q WLD} = 105\,\mathrm{s}^{-1}$ and $\Omega_0 = \sqrt{\frac{k}{m}} = \sqrt{\frac{ED^2}{0.972\rho_q L^4}} = 2.27 \cdot 10^5\,\frac{\mathrm{rad}}{\mathrm{s}}$.

(f) By replacing $x = x_0\cos(\Omega t + \phi)$ in $\ddot{x} + \gamma\dot{x} + \Omega_0^2 x = f_0\cos(\Omega t + \phi - \delta\varphi)$, where $\delta\varphi = \phi - \varphi$ and $f_0 = F_0/m$, we obtain $-\Omega^2 x_0\cos(\Omega t + \phi) - \gamma\Omega x_0\sin(\Omega t + \phi) + \Omega_0^2 x_0\cos(\Omega t + \phi) = f_0\cos(\Omega t + \phi)\cos\delta\varphi + f_0\sin(\Omega t + \phi)\sin\delta\varphi$. After identification of the terms in front of $\sin(\Omega t + \phi)$ and $\cos(\Omega t + \phi)$ we obtain the system of equations $-x_0(\Omega^2 - \Omega_0^2) = f_0\cos\delta\varphi$ and $-\gamma\Omega x_0 = f_0\sin\delta\varphi$, the solutions to which are $x_0 = \frac{f_0}{\sqrt{(\Omega^2-\Omega_0^2)^2 + \gamma^2\Omega^2}}$ and $\delta\varphi = \mathrm{arctg}(\frac{\gamma\Omega}{\Omega^2-\Omega_0^2})$.

(g) The amplitude of electrical power that the voltage source gives to the tines of the fork is $P = \eta\dot{x}V_0$. This amplitude corresponds to the amplitude of mechanical power of the driving force that acts on the tines of the fork. Consequently we have $P = 2F_0\dot{x}$, and therefore $\eta V_0 = 2F_0$.

(h) Using the relation between the charge and the elongation of the top of the tine, the equation of oscillations of the charge reads $\frac{m}{\eta}\ddot{q} + \frac{b}{\eta}\dot{q} + \frac{k}{\eta}q = F_0\cos(\Omega t + \varphi) = \frac{\eta}{2}V_0\cos(\Omega t + \varphi)$. By reducing the previous equation to the form

$L_e\ddot{q} + R_e\dot{q} + \frac{1}{C_e}q = V_0\cos(\Omega t + \varphi)$, it follows that $L_e = \frac{2m}{\eta^2} = \frac{0.486\rho_q WLD}{\eta^2} =$
$2.27\,\text{kH}$, $R_e = \frac{2b}{\eta^2} = 237\,\text{k}\Omega$ and $C_e = \frac{\eta^2}{2k} = 8.54\,\text{fF}$.

(i) An equivalent electrical circuit consists of a serial $R_e L_e C_e$ circuit that represents the mechanical part of the quartz fork and the capacitor C connected in parallel to this part of circuit that represents the electrodes. An equivalent circuit is shown in Figure 5.10.

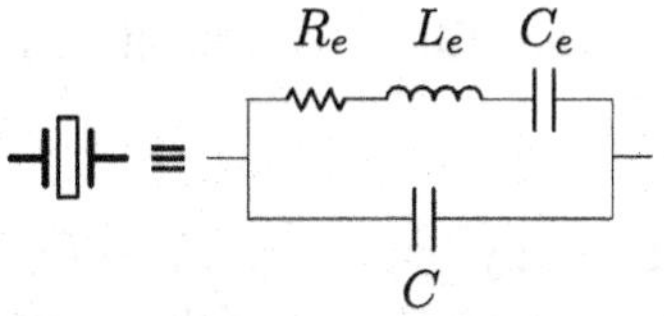

Figure 5.10 The equivalent electrical circuit of a piezoelectric oscillator. The electrical symbol for the piezoelectric oscillator is shown on the left, while the corresponding equivalent circuit is shown on the right.

(j) The quartz clock shows the correct time when $\frac{\Delta f}{f} = 0$. From the graph, we see that this is the case at a temperature $t_0 = 25\,°\text{C}$. Since $f \sim 1/T$, we have $\frac{\Delta f}{f} = \frac{\Delta T}{T} < 0$ in both cases. This means that both clocks are late.

(k) Since it was stated in the problem that the dependence is parabolic and we see from the graph that the two zeros are equal to $t_0 = 25\,°\text{C}$, we have that $\frac{\Delta f}{f} = \alpha(t - t_0)^2$, where α is a constant that has to be determined. By choosing one point from the graph, for example $(-70 \cdot 10^{-6}, -20\,°\text{C})$, we obtain $\alpha = \frac{-70\cdot10^{-6}}{2025\,°\text{C}^2} = -3.5 \cdot 10^{-8}\,°\text{C}^{-2}$. Consequently, the dependence of frequency on temperature is $f(t) = (1 - 3.5 \cdot 10^{-8}(t/°\text{C} - 25)^2) \cdot 2^{15}\,\text{Hz}$.

(l) The clock in Vladimir's office will be accurate after a year, while the clock in Nenad's office will be late by ΔT. Since $\frac{\Delta f}{f} = \frac{\Delta T}{T}$, it follows that $\Delta T = \frac{\Delta f}{f}T = -3.5 \cdot 10^{-8}\,°\text{C}^{-2}(10\,°\text{C} - 25\,°\text{C})^2 \cdot 365 \cdot 24 \cdot 60\,\text{min} \approx -4.14\,\text{min}$.

Problem 36   Solar Cells

How is solar energy converted to electricity?
What is the origin of energy losses in a solar cell?
When is a solar cell most efficient?

Solar cells are devices that convert the energy of the electromagnetic radiation of the sun into electrical energy. Solar cells are typically made of semiconducting materials.

It is known from the band theory of solids that the energy of an electron in the material can take values only from certain allowed intervals of energy – so-called bands. Other energies are forbidden and these regions of energy are called energy gaps. The bands that are of relevance for the description of electrical and optical properties of materials are the valence band that covers the interval of energies $(E_V - W_V, E_V)$ and the conduction band from the interval of energies $(E_C, E_C + W_C)$, where $E_C > E_V$. These two bands are separated by an energy gap in the region of energies (E_V, E_C). In the case of a semiconducting material at zero temperature all the states in the valence band are filled with electrons, while all the states in the conduction band are empty.

(a) Calculate the values of energy gap $E_g = E_C - E_V$ that allow the material to absorb all photons from the visible part of the spectrum $\lambda \in (400, 700)$ nm.

At higher temperatures T, some of the electrons from the valence band can make a transition to the conduction band. The probability that the electronic state of energy E is filled is given by the Fermi–Dirac function $f_{FD}(E) = \frac{1}{e^{\frac{E-E_F}{k_B T}} + 1}$, where the quantity E_F is called the Fermi energy. When an electron from the valence band makes a transition to the conduction band, it leaves an empty state in the valence band – a so-called hole.

The density of states is defined as $D(E) = \frac{dn_s}{dE}$, where dn_s is the number of allowed states in the small interval of energies $(E, E + dE)$ in unit volume of the material. Assume that the density of states in the interval $(E_C, E_C + W_C)$ (conduction band) is given as $D(E) = D_C$, while it is given as $D(E) = D_V$ in the interval $(E_V - W_V, E_V)$ (valence band).

(b) Derive the expression for the total concentration of electrons in the conduction band $n_e^{(0)}$ in terms of E_C, W_C, E_F, T, and D_C. Next simplify the expression using the approximations $E_C - E_F \gg k_B T$ and $W_C \gg k_B T$.

(c) Derive the expression for the total concentration of holes in the valence band $n_p^{(0)}$ in terms of E_V, W_V, E_F, T, and D_V. Then simplify the expression using that $E_F - E_V \gg k_B T$ and $W_V \gg k_B T$.

Concentration of electrons and holes in the semiconductor can be controlled by changing the position of Fermi energy (which can be technologically achieved, for example, by doping the semiconductor – adding additional atoms in the material). A semiconductor with the concentration of electrons much larger than the concentration of holes is a semiconductor of n-type, while a semiconductor with a concentration of holes much larger than the concentration of electrons is a semiconductor of p-type.

(d) Determine the Fermi energy for which the concentration of electrons is at least ten times larger than the concentration of holes (i.e., it is an n-type semiconductor), as well as the Fermi energy for which the concentration of holes is at least ten times larger than the concentration of electrons (i.e., the semiconductor is of p-type). Express your results in terms of E_C, E_V, $k_B T$, and the ratio D_C/D_V.

A semiconductor solar cell is usually made of a junction of an n-type and a p-type semiconductor, which are both based on the same material, typically silicon. Fermi energy must be equal in both semiconductors that are in contact in the case of thermodynamic equilibrium. This leads to differences of energies of the bottom of conduction band $E_C^{(p)} - E_C^{(n)}$ and the energies of the top of the valence band $E_V^{(p)} - E_V^{(n)}$ in the two materials. In the same time the energy gap of each of the two materials does not change.

(e) Determine $E_C^{(p)} - E_C^{(n)}$ and $E_V^{(p)} - E_V^{(n)}$. Express your result in terms of the temperature, the concentration of electrons $n_e^{(0)}$ and holes $n_p^{(0)}$ in a semiconductor that is not doped, the concentration of electrons n_e in the semiconductor of n-type, and the concentration of holes n_p in the semiconductor of p-type. Show that the quantities in question are positive.

We assume that the sun radiates as a black body. The number of photons emitted by a black body in unit time per unit surface of the body in the interval of frequencies $(v, v + dv)$ is

$$\frac{dN_v}{dt\, dS} = \frac{2\pi}{c^2} \frac{v^2 dv}{\exp\left(\frac{hv}{k_B T}\right) - 1}. \tag{5.16}$$

The temperature of the sun is T_S, while the area of the surface of the sun is A_S. We consider a solar cell in the shape of a thin plate. The area of one side of the plate is A_p. The plate is at a distance L from the sun, while the angle between the line perpendicular to the plate and the direction that connects the plate and the sun is θ_{ps}, as shown in Figure 5.11.

(f) Determine the number of photons emitted from the sun that is incident on the plate in unit time. You should not solve the integral that appears in the solution.

Next we assume that each photon that is incident on the solar cell gets absorbed if its energy is larger than the energy gap. An electron–hole pair is created after absorption. The loss of energy occurs because electrons make a transition to the bottom of the conduction band due to interaction with the lattice, while the holes make a transition to the top of the valence band for the same reason. Consequently the energy of the electron–hole pair almost instantaneously decreases from the incident photon energy to the energy of the gap.

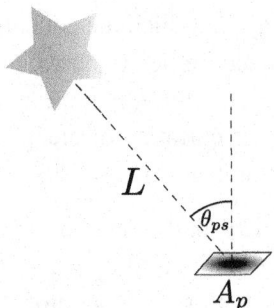

Figure 5.11 Spatial position of a solar cell

(g) Determine the efficiency of the solar cell assuming that the only source of energy loss is the process described in the previous paragraph. The efficiency of the solar cell is defined as the ratio of the energy produced and the total energy of the photons incident on the solar cell. Express the final result in terms of the quantity $x_g = \frac{E_g}{k_B T_S}$. Do not attempt to solve the integrals that appear in this part of the problem.

(h) Plot the graph of the dependence of the solar cell efficiency on x_g. The graph of the function $g(x_g) = \int_{x_g}^{\infty} \frac{x^2 \, dx}{e^x - 1}$ is given in Figure 5.12, while $\int_0^{\infty} \frac{x^3 \, dx}{e^x - 1} = \frac{\pi^4}{15}$.

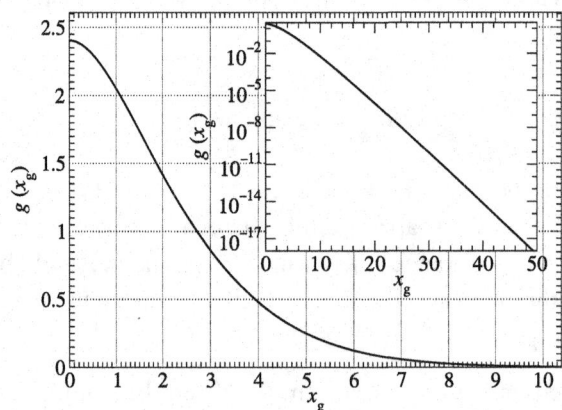

Figure 5.12 The graph of the function $g(x_g)$ in linear and logarithmic scale

(i) Determine the maximal efficiency of the solar cell and the value of x_g for which this maximum is achieved. What is the value of the energy gap when this efficiency is reached if the temperature of the sun is $T_S = 5.8 \cdot 10^3 \, K$?

Another type of energy loss in the solar cell comes from the radiative recombination of generated electrons and holes. In radiative recombination an electron from the conduction band makes a transition to the valence band and emits a photon. This process can be effectively thought of as the process where an electron and a hole disappear while a photon is created. To determine the number of recombined electron–hole pairs, we first consider the following problem.

(j) Consider a solar cell located in a closed cavity. The temperature of the walls of the cavity is T_c and the walls emit radiation as a black body. Determine the number of photons absorbed by the solar cell in unit time. For an ideal gas of photons, the number of photons in unit volume in the interval of frequencies $(\nu, \nu + d\nu)$ is

$$dn_\nu = \frac{8\pi}{c^3} \frac{\nu^2 d\nu}{\exp\left(\frac{h\nu}{k_B T}\right) - 1}.$$

Do not attempt to solve the integral over frequencies that appears in the solution.

(k) Determine the number of recombination events in unit time F_0 when the solar cell is in thermodynamic equilibrium at a temperature T_c and it is not exposed to the radiation of the sun.

Next we consider a solar cell exposed to the radiation of the sun. Since $E_C^{(p)} - E_C^{(n)} > 0$ and $E_V^{(p)} - E_V^{(n)} > 0$ the electrons generated in p-region move to n-region, while the holes generated in n-region move to p-region. We assume that energy distribution of electrons and holes in this nonequilibrium state can be described using the Fermi–Dirac function where the Fermi energies for electrons $E_F^{(n)}$ and holes $E_F^{(p)}$ are different. The difference of these energies is related to the applied voltage U as $E_F^{(n)} - E_F^{(p)} = eU$.

To determine the number of radiative recombination events in the solar cell exposed to the sun, we assume that the number of recombination events in unit time is proportional to the product of concentrations of generated electrons and holes.

(l) Determine the ratio of the number of recombination events in unit time F in the solar cell exposed to the sun and the corresponding quantity in equilibrium F_0. Express the result in terms of the applied voltage on the cell exposed to the sun and the temperature of the cell. Assume in this and the following parts of the problem that the cell absorbs only the photons that arrive from the sun.

(m) Determine the voltage U_{oc} on a cell that is exposed to the sun but is not connected to an external circuit. Calculate its numerical value if the temperature of the solar cell is $T_c = 300\,\mathrm{K}$, the energy gap is $E_g = 1.1\,\mathrm{eV}$, $\theta_{ps} = 0$, the distance between the Earth and the sun is $L = 1.5 \cdot 10^{11}\,\mathrm{m}$, and the surface area of the sun is $A_S = 6.16 \cdot 10^{18}\,\mathrm{m}^2$.

(n) Determine the dependence of the current I through the solar cell on the voltage at its ends $U \in (0, U_{oc})$ and plot the corresponding graph. Use the same numerical values as in the previous part of the problem and assume that the area of one side of the solar cell is $A_p = 1.0 \, \text{cm}^2$.

(o) What values of I and U maximize the power that the cell gives to the external circuit?

(p) What is the efficiency of the solar cell for these values?

Planck's constant is $h = 6.63 \cdot 10^{-34} \, \text{J} \cdot \text{s}$, the speed of light in vacuum is $c = 3.00 \cdot 10^8 \, \text{ms}^{-1}$, the elementary charge is $e = 1.60 \cdot 10^{-19} \, \text{C}$, and the Boltzmann constant is $k_B = 8.62 \cdot 10^{-5} \, \frac{\text{eV}}{\text{K}}$.

Solution of Problem 36

(a) The photon is absorbed when its energy is larger than the energy gap. Consequently $\frac{hc}{\lambda} > E_g$, which leads to $E_g < \frac{hc}{\lambda_1}$ and $E_g < \frac{hc}{\lambda_2}$, where $\lambda_1 = 400$ nm and $\lambda_2 = 700$ nm. Previous inequalities imply that the energy gap should satisfy the condition $E_g < \frac{hc}{\lambda_2} = 1.78 \, \text{eV}$.

(b) The number of allowed states in unit volume of the material in the interval of energies $(E, E + dE)$ is $D_C(E)dE$. Since the probability of occupation of the state by an electron is $f_{FD}(E) = \frac{1}{\exp \frac{E-E_F}{k_B T} + 1}$, the concentration of electrons whose energies are from this interval is $dn_e^{(0)} = f_{FD}(E)D_C(E)dE$. Total concentration of conduction band electrons is obtained by integration of the previous equation through the interval of conduction band energies $n_e^{(0)} = \int_{E_C}^{E_C + W_C} \frac{1}{\exp \frac{E-E_F}{k_B T} + 1} D_C(E)dE$. By neglecting the term 1 in the denominator (this is allowed because $\frac{E-E_F}{k_B T} \gg 1$ for the energies E from the conduction band), we obtain $n_e^{(0)} = D_C \int_{E_C}^{E_C + W_C} \exp\left(-\frac{E-E_F}{k_B T}\right) dE$. After solving the integral we find

$$n_e^{(0)} = D_C k_B T \exp\left(-\frac{E_C - E_F}{k_B T}\right)\left[1 - \exp\left(-\frac{W_C}{k_B T}\right)\right].$$

Since $\frac{W_C}{k_B T} \gg 1$, the last expression can be approximated as $n_e^{(0)} = D_C k_B T \exp\left(-\frac{E_C - E_F}{k_B T}\right)$.

(c) The probability that the state of energy E in the valence band is empty is $1 - f_{FD}(E)$. In a similar manner as in part (b) we find

$$n_p^{(0)} = \int_{E_V - W_V}^{E_V} \left(1 - \frac{1}{\exp \frac{E-E_F}{k_B T} + 1}\right) D_V(E)dE, \text{ which can be also written as}$$

$n_p^{(0)} = \int_{E_V-W_V}^{E_V} \frac{1}{\exp\frac{E_F-E}{k_BT}+1} D_V(E)dE$. Using similar approximations as in part (b)

we obtain $n_p^{(0)} = D_V k_B T \exp\left(-\frac{E_F-E_V}{k_BT}\right)$.

(d) Using the solutions of parts (b) and (c), we find that $n_e^{(0)} > 10 \cdot n_p^{(0)}$ when

$\frac{D_C}{D_V} e^{\frac{2E_F-E_C-E_V}{k_BT}} > 10$. Consequently the semiconductor is of n-type when $E_F >$

$\frac{E_C+E_V}{2} + \frac{1}{2}k_BT\ln\left(\frac{10D_V}{D_C}\right)$. The semiconductor is of p-type when $n_p^{(0)} > 10\cdot n_e^{(0)}$,

which leads to $E_F < \frac{E_C+E_V}{2} - \frac{1}{2}k_BT\ln\left(\frac{10D_C}{D_V}\right)$.

(e) After multiplication of the expressions from the solutions of parts (b) and (c)
we obtain

$$n_e^{(0)}n_p^{(0)} = D_C D_V (k_B T)^2 \exp\left(-\frac{E_g}{k_BT}\right), \tag{5.17}$$

where $E_g = E_C - E_V$ is the energy gap. We next have $n_e = D_C k_B T \exp\left(-\frac{E_C^{(n)}-E_F}{k_BT}\right)$

and $n_p = D_V k_B T \exp\left(-\frac{E_F-E_V^{(p)}}{k_BT}\right)$, as well as $E_V^{(n)} = E_C^{(n)} - E_g$ and $E_V^{(p)} + E_g =$

$E_C^{(p)}$. Previous equations lead to

$$E_C^{(p)} - E_C^{(n)} = E_V^{(p)} - E_V^{(n)} = k_B T \ln\left(\frac{n_e n_p}{n_e^{(0)}n_p^{(0)}}\right).$$

Since $n_e > n_e^{(0)}$ and $n_p > n_p^{(0)}$, the quantities in question are positive.

(f) The total number of photons emitted by the sun in unit time is $N_S' = A_S \frac{2\pi}{c^2} \int_0^\infty \frac{v^2 dv}{\exp\left(\frac{hv}{k_BT_S}\right)-1}$. The photons that are incident on the solar cell are those

that are emitted to the solid angle subtended by the cell to an observer from
the sun. The fraction of these photons in the total number of photons emitted
from the sun is $f = \frac{A_p\cos\theta_{ps}}{4\pi L^2}$, where $A_p\cos\theta_{ps}$ is the area of the projection of
the cell on the sphere of radius L whose center is in the sun, while $4\pi L^2$ is the
surface area of this sphere. The number of photons incident on the cell in unit
time is therefore

$$N_p' = fN_S' = \frac{A_p\cos\theta_{ps}}{4\pi L^2}A_S\frac{2\pi}{c^2}\int_0^\infty \frac{v^2 dv}{\exp\left(\frac{hv}{k_BT_S}\right)-1}. \tag{5.18}$$

(g) The number of photons incident on the cell in unit time is given by equation
(5.18), while the total energy of these photons is obtained when the function
under the integral in this expression is multiplied by the energy of one photon

$h\nu$. Therefore, the energy of all photons that are incident on the cell in unit time is

$$E'_p = \frac{A_p \cos\theta_{ps}}{4\pi L^2} A_S \frac{2\pi}{c^2} \int_0^\infty \frac{h\nu \cdot \nu^2 d\nu}{\exp\left(\frac{h\nu}{k_B T_S}\right) - 1}. \tag{5.19}$$

Only photons of energy larger than E_g contribute to useful energy, while the contribution of each of these photons to useful energy is exactly E_g. Consequently the useful energy stemming from photons that are incident on the cell in unit time is

$$E'_k = \frac{A_p \cos\theta_{ps}}{4\pi L^2} A_S \frac{2\pi}{c^2} \int_{E_g/h}^\infty \frac{E_g \nu^2 d\nu}{\exp\left(\frac{h\nu}{k_B T_S}\right) - 1}. \tag{5.20}$$

The efficiency of the solar cell is then $\eta = \frac{E'_k}{E'_p}$, which leads to

$$\eta = \frac{\int_{E_g/h}^\infty \frac{E_g \nu^2 d\nu}{\exp\left(\frac{h\nu}{k_B T_S}\right) - 1}}{\int_0^\infty \frac{h\nu^3 d\nu}{\exp\left(\frac{h\nu}{k_B T_S}\right) - 1}}. \tag{5.21}$$

After the replacements $x = \frac{h\nu}{k_B T_S}$ and $x_g = \frac{E_g}{k_B T_S}$, the last expression reduces to

$$\eta = \frac{x_g \int_{x_g}^\infty \frac{x^2 \, dx}{e^x - 1}}{\int_0^\infty \frac{x^3 \, dx}{e^x - 1}}.$$

(h) From the definition of the function $g(x_g)$ and the integral given in the text of the problem, we find $\eta = \frac{15 x_g}{\pi^4} g(x_g)$. The graph of the dependence of the solar cell efficiency on x_g is shown in Figure 5.13.

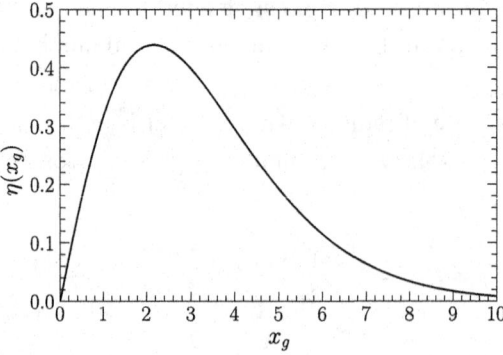

Figure 5.13 The graph of the dependence of solar cell efficiency on x_g

(i) We read from the graph shown in Figure 5.13 that maximal efficiency is $\eta_{max} = 0.44$ and it is reached for $x_g^{(m)} = 2.2$. Optimal energy gap of the material is therefore $E_g^{(m)} = k_B T_S x_g^{(m)} = 1.1\,\text{eV}$.

(j) Since the cell is inside the cavity that emits as a black body, photons in the cavity constitute an ideal gas. The concentration of photons the cell can absorb is given as $n = \int_{v=E_g/h}^{\infty} dn_v$. Among the photons whose velocity is directed at an angle from the small interval $(\theta, \theta + d\theta)$, the photons that are located in the volume $dV = A_p c \cos\theta\, dt$ shown in Figure 5.15 are incident on one surface of the cell in small time dt. The fraction of photons whose velocity is directed at an angle from small interval $(\theta, \theta + d\theta)$ in the total number of photons is equal to the ratio of the area of the ring on the sphere shown in Figure 5.14 and the total area of that sphere $du = \frac{2\pi \sin\theta\, d\theta}{4\pi} = \frac{1}{2}\sin\theta\, d\theta$. Consequently the total number of photons incident on one side of the cell in time dt that is absorbed is $\int_{\theta=0}^{\pi/2} du \cdot n \cdot dV$. Therefore, the total number of photons absorbed by the cell in unit time is

$$N_j' = 2 \int_{\theta=0}^{\pi/2} \frac{1}{2} \sin\theta \cdot d\theta \cdot n A_p c \cos\theta,$$

where the factor 2 originates from the fact that the cell has two surfaces. Since $\int_{\theta=0}^{\pi/2} \sin\theta \cos\theta\, d\theta = \frac{1}{2}$, we obtain from previous expressions that

$$N_j' = \frac{4\pi}{c^2} A_p \int_{v=E_g/h}^{\infty} \frac{v^2 dv}{\exp\left(\frac{hv}{k_B T_c}\right) - 1}. \tag{5.22}$$

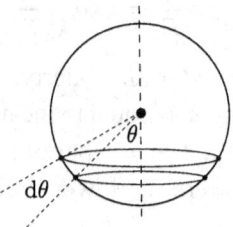

Figure 5.14 The ring on the sphere that defines the region of space defined by the angle from small interval $(\theta, \theta + d\theta)$

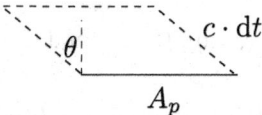

Figure 5.15 The volume $dV = A_p c \cos\theta\, dt$ occupied by the photons that are incident on the cell

(k) We obtained in part (j) the number of photons absorbed by the cell in unit time (which is at the same time the number of generated electron–hole pairs in unit time), when the cell is in thermodynamic equilibrium at a temperature T_c. The number of generated electron–hole pairs in thermodynamic equilibrium must be equal to the number of recombined electron–hole pairs. Therefore, the number of recombination events in unit time is equal to

$$F_0 = N'_j = \frac{4\pi}{c^2} A_p \int_{\nu=E_g/h}^{\infty} \frac{\nu^2 d\nu}{\exp\left(\frac{h\nu}{k_B T_c}\right) - 1}.$$ (5.23)

(l) Since the number of recombination events is proportional to the product of the number of electrons and holes, the ratio in question is $\frac{F}{F_0} = \frac{n_e n_p}{n_e^{(0)} n_p^{(0)}}$. It follows from the solution of part (b) that $\frac{n_e}{n_e^{(0)}} = \exp \frac{E_F^{(n)} - E_F}{k_B T_c}$, while it follows from the solution of part (c) that $\frac{n_p}{n_p^{(0)}} = \exp\left(-\frac{E_F^{(p)} - E_F}{k_B T_c}\right)$, where E_F is the Fermi energy when the cell is not exposed to the radiation of the sun. We obtain from previous equations that

$$\frac{F}{F_0} = \exp \frac{E_F^{(n)} - E_F^{(p)}}{k_B T_c}.$$ (5.24)

Using the relation between the difference of Fermi energies and the voltage given in the text of the problem, we obtain

$$\frac{F}{F_0} = \exp \frac{eU}{k_B T_c}.$$ (5.25)

(m) When the cell is not connected to the external circuit, the number of photons absorbed by the cell in unit time is equal to the number of recombination events in unit time – that is, $F = N'_a$. The number of photons absorbed in unit time is given by equation (5.18) where the lower limit of integration is replaced by E_g/h – that is,

$$N'_a = \frac{A_p \cos \theta_{ps}}{4\pi L^2} A_S \frac{2\pi}{c^2} \int_{E_g/h}^{\infty} \frac{\nu^2 d\nu}{\exp\left(\frac{h\nu}{k_B T_S}\right) - 1}.$$ (5.26)

We obtain from equations (5.26), (5.25), and (5.23) that

$$U_{oc} = \frac{k_B T_c}{e} \ln \left[\frac{A_s \cos \theta_{ps}}{8\pi L^2} \left(\frac{T_S}{T_c}\right)^3 \frac{g\left(\frac{E_g}{k_B T_S}\right)}{g\left(\frac{E_g}{k_B T_c}\right)} \right].$$ (5.27)

We finally obtain the numerical value $U_{oc} = 0.85$ V.

(n) The current in external circuit originates from all generated electron–hole pairs, except those that recombine before they reach the contacts. We therefore have $I = e\,(N'_a - F)$. Using equations (5.26), (5.25), and (5.23), we find that the dependence of the current on voltage is

$$I = a - b \exp\frac{U}{U_c}, \qquad (5.28)$$

where $U_c = \frac{k_B T_c}{e}$, $a = \frac{eA_p A_S \cos\theta_{ps}}{2L^2 c^2}\left(\frac{k_B T_S}{h}\right)^3 g\left(\frac{E_g}{k_B T_S}\right)$, and $b = \frac{eA_p \cdot 4\pi}{c^2}\left(\frac{k_B T_c}{h}\right)^3 g\left(\frac{E_g}{k_B T_c}\right)$. We then find $a = 55\,\text{mA}$ and $b = 3.5 \cdot 10^{-16}\,\text{A}$. The graph in question is given in Figure 5.16.

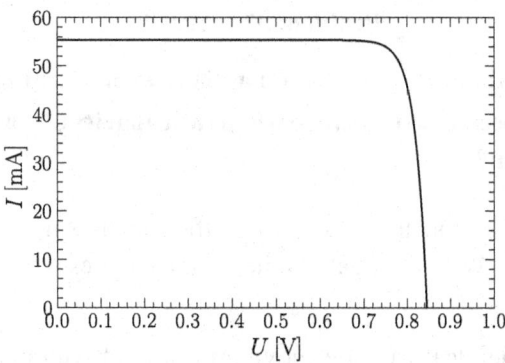

Figure 5.16 Current-voltage characteristic of the solar cell

(o) The power the solar cell gives to the external circuit is $P_k = UI = U\left(a - b\exp\frac{U}{U_c}\right)$. The derivative of the function $P_k(U)$ is $\frac{dP_k}{dU} = a - b\left(1 + \frac{U}{U_c}\right)\exp\frac{U}{U_c}$. We numerically solve the equation $a - b\left(1 + \frac{U_m}{U_c}\right)\exp\frac{U_m}{U_c} = 0$ and find that the derivative is zero when $U_m = 0.76\,\text{V}$ (where it is positive for $U < U_m$, while it is negative for $U > U_m$). The power is therefore maximal for $U = U_m$. The current is then equal to $I_m = 54\,\text{mA}$.

(p) The efficiency of the solar cell is given as $\eta = \frac{P_k}{E'_p}$, where E'_p is given by equation (5.19). By introducing the replacement $x = \frac{h\nu}{k_B T_S}$ and using $\int_0^\infty \frac{x^3\,dx}{e^x - 1} = \frac{\pi^4}{15}$ we obtain $E'_p = \frac{A_p A_S \cos\theta_{ps}}{2L^2 c^2}\frac{(k_B T_S)^4}{h^3}\frac{\pi^4}{15} = 139\,\text{mW}$. The efficiency of the solar cell is then $\eta = \frac{U_m I_m}{E'_p} = 29\,\%$.

We refer readers interested in the physics describing the efficiency limits of solar cells to reference [31].

6

Miscellaneous

Problem 37 Traffic

What is the maximal flow of cars on a highway in a traffic jam?

How does the speed of cars change when the capacity of a highway is reduced due to construction?

The motion of cars on the road in a traffic jam is similar to the motion of compressible fluid. We investigate some characteristics of such a fluid in this problem.

(a) Assume that identical cars are moving on a road with one traffic lane. The length of each car is l. The driver of each car keeps sufficient distance from the car in front to avoid collision even when the car in front suddenly stops. The magnitude of deceleration when the driver brakes is a, while the time it takes for the driver to start braking is t_r. Determine the dependence of the number of cars per unit length of the road n (which is analogous to the fluid density) on the car speed v (and quantities a, l, and t_r).

(b) Determine the dependence of the flow of cars q (the number of cars passing in unit time through the cross section of the road) on the car speed v (and quantities a, l, and t_r).

(c) Find the value of v that maximizes the flow q.

(d) Calculate the maximal flow if $a = 10.0\,\frac{\text{m}}{\text{s}^2}$, $t_r = 1.00$ s, and $l = 5.00$ m. Calculate also the speed that maximizes the flow.

(e) What is the maximal flow of cars on a highway with three traffic lanes?

(f) Due to construction, on a part of the highway with three traffic lanes only one lane is open. How many times will the maximal flow be smaller in comparison to when all traffic lanes are open?

(g) How many times will the car speed (on a part of the highway where all three traffic lanes are open, which is located before the part with construction) be smaller in the case described in part (f) in comparison to when all three traffic lanes are open? Assume that the cars are moving in such a way to maximize the flow.

Solution of Problem 37

(a) The distance between the front of the car and the rear of the car in front is $s = s_1 + s_2$, where $s_1 = vt_r$ is the distance the car travels before the driver starts to brake, while $s_2 = \frac{v^2}{2a}$ is the distance the car travels during braking. The number of cars per unit length of the road is then $n = \frac{1}{s+l}$, which leads to

$$n = \frac{2a}{v^2 + 2avt_r + 2al}. \tag{6.1}$$

(b) The flow is $q = nv$ and consequently

$$q = \frac{2av}{v^2 + 2avt_r + 2al}. \tag{6.2}$$

(c) The expression for the dependence of flow on speed can be presented in the form

$$q = \frac{2a}{v + \frac{2al}{v} + 2at_r}. \tag{6.3}$$

The flow will be maximal when the function $f(v) = v + \frac{2al}{v}$ exhibits a minimum. The first derivative of this function is $f'(v) = 1 - \frac{2al}{v^2}$. From this expression we see that $f'(v) = 0$ for $v = v_m = \sqrt{2al}$, while $f'(v) > 0$ for $v > v_m$ and $f'(v) < 0$ for $v < v_m$. Consequently $f(v)$ is minimal when $v = v_m = \sqrt{2al}$.

(d) By replacing $v = \sqrt{2al}$ in equation (6.3) we find that the maximal flow is $q_m = \frac{a}{at_r + \sqrt{2al}} = 0.500\,\mathrm{s}^{-1}$. It is obtained when the speed is $v_m = \sqrt{2al} = 10.0\,\frac{\mathrm{m}}{\mathrm{s}}$.

(e) The maximal flow in each of the traffic lanes is q_m. The maximal flow on the highway with three traffic lanes is then $q_3 = 3q_m = 1.50\,\mathrm{s}^{-1}$.

(f) When only one lane is open, the maximal flow in that lane is $q_1 = q_m = 0.500\,\mathrm{s}^{-1}$, which is also the maximal flow on the highway. This implies that the maximal flow is three times smaller than when all lanes are open.

(g) We denote as u the speed of the car on the part of the road with three traffic lanes. The flow on that part of the road is on one hand $3q(u)$, while on the other hand it is q_1, which implies

$$q_1 = 3\frac{2au}{u^2 + 2aut_r + 2al}. \tag{6.4}$$

By solving equation (6.4) with respect to u we obtain

$$u_{1/2} = a\left(\frac{3}{q_1} - t_r\right) \pm \sqrt{a^2\left(\frac{3}{q_1} - t_r\right)^2 - 2al}. \tag{6.5}$$

The numerical values of the solutions are $u_1 = 99.0 \frac{m}{s}$ and $u_2 = 1.01 \frac{m}{s}$. The solution u_1 is outside the range of speeds that cars can reach on the highway. Consequently the only acceptable solution is $u = u_2 = 1.01 \frac{m}{s}$. This implies that the speed of the car will be $r = \frac{v_m}{u} = 9.90$ times smaller.

The authors presented this problem at the national physics competition for the fourth grade of high school in Serbia in 2013.

Problem 38   Mexican Wave

Can a dance of spectators be described using mechanical waves?
How do local interactions of spectators produce a global single wave?

A Mexican wave is a phenomenon at major sport events that occurs when spectators stand up raising their arms and then sit down again. The name originates from the 1986 World Cup in football held in Mexico when such an audience gesture was first popularized worldwide. Since then, the Mexican wave appears in all cultures and sports, while it is called "la ola" in Brazil, Germany, Italy, and some other countries.

Assume that all spectators move identically and that the height of each spectator is $h_1 = 1.1\,\text{m}$ in a sitting position and $h_2 = 2.3\,\text{m}$ in a standing position with arms raised, as shown in Figure 6.1. The person at $x = 0\,\text{m}$ starts the Mexican wave at $t = 0$ s. The spectator at his/her left side stands up after time interval Δt and moves in an identical manner and so on. When a spectator sits down, he/she stands up again and continues periodically to sit down and stand up.

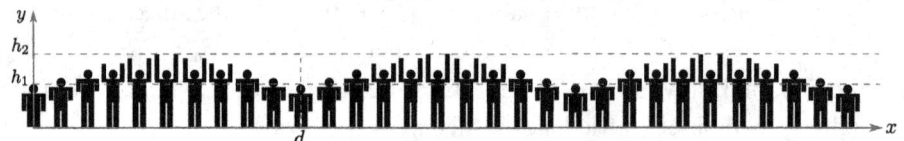

Figure 6.1 Mexican wave

From the analysis of videos of different sport events, one can conclude that the width of the wave (see Figure 6.1) is $d = 6\,\text{m}$, which corresponds to $n = 10$ seats. The time it takes for a spectator to stand up, raise his/her arms, and sit down again is $T = 1\,\text{s}$. From the point of view of physics, the Mexican wave shown in Figure 6.1 can be described using the equation of a mechanical wave $y(x,t) = B + A\cos(\omega t + kx + \varphi)$, where y is the height of the wave at the position x and time t.

(a) Calculate the constant B, the amplitude A, the wave vector k, the angular frequency ω, and the phase φ of the Mexican wave.

(b) Calculate the speed of the Mexican wave and the time interval Δt.
(c) The speed of mechanical waves is usually larger than the speed of particles that form the wave. Is this also the case for the Mexican wave?

At less popular sport events, the audience is not so excited to stand up and sit down all the time, forming the Mexican wave shown in Figure 6.1. Instead they form a wave pulse that travels through the whole stadium or at least at its part. One such pulse at the moment of formation (at time $t_0 = 0\,\text{s}$) is shown in Figure 6.2.

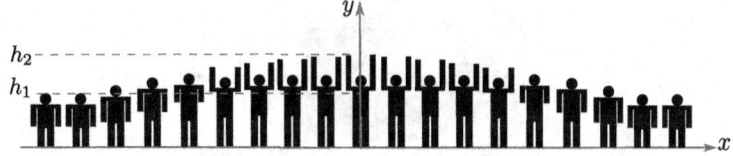

Figure 6.2 The pulse of the Mexican wave

It was noted that the pulse has a parabolic shape that does not change in time. The pulse travels at a constant speed $u = 12\,\frac{\text{m}}{\text{s}}$ in the direction of the x-axis. The width of the wave is $d_p = 9\,\text{m}$, which corresponds to $n_p = 15$ seats.

(d) Determine the equation that describes the pulse $y_p(x,t)$.
(e) Calculate the time interval Δt in this case.
(f) Plot the pulse at times $t_1 = 2\,\text{s}$, $t_2 = 4\,\text{s}$, and $t_3 = 6\,\text{s}$.
(g) Plot the graph of the time dependence of the speed of the highest point of the spectator's body for the spectators at $x = 0\,\text{m}$, $x = 4.5\,\text{m}$ from time $t_0 = 0\,\text{s}$ to time $t_4 = 1\,\text{s}$.
(h) What would be the equation of the pulse $y_p(x,t)$ if it were traveling in the direction opposite to the x-axis?

Consider the motion of the pulse of the Mexican wave at the stadium shown in Figure 6.3. The part of the stadium intended for spectators is in the shape of an ellipsoidal ring. The lengths of half-axes of the inner ellipse are a_1 and b_1 while they are a_2 and b_2 for the outer ellipse. The pulse of the Mexican wave propagates in each row. The wave front of the pulse has the direction of the radius vector $\vec{r}$ at all times. The equation of the ellipse with the length of half-axes a and b in cylindrical coordinates (r, φ) is given as

$$r(\varphi) = \frac{ab}{\sqrt{(b\cos\varphi)^2 + (a\sin\varphi)^2}},$$

while the circumference of the ellipse can be approximated as

$$O \approx \pi[3(a+b) - \sqrt{(3a+b)(a+3b)}].$$

(i) The speed of the Mexican wave in the first row (the inner ellipse) propagates at a constant speed u_1. Determine the speed $u_2(\varphi)$ of the pulse propagating in the last row (the outer ellipse).

(j) What is the mean value of the speed $u_2(\varphi)$ when the pulse travels through the whole stadium?

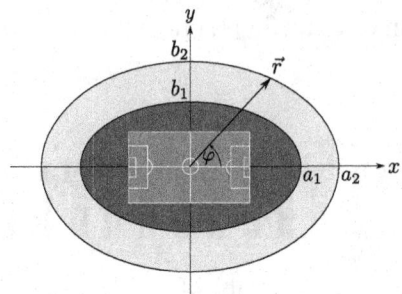

Figure 6.3 The sketch of the football stadium

Solution of Problem 38

(a) The amplitude of the Mexican wave is $A = \frac{1}{2}(h_2 - h_1) = 0.6\,\text{m}$, the constant $B = \frac{1}{2}(h_1 + h_2) = 1.7\,\text{m}$, the wave vector is $k = \frac{2\pi}{d} = \frac{2\pi}{\lambda} = 1.05\,\text{m}^{-1}$, while the angular frequency is $\omega = \frac{2\pi}{T} = 6.28\,\text{rad/s}$. The phase can be determined from the condition $y(0,0) = B + A\cos\varphi = h_1$, which leads to $\varphi = \pi$.

(b) The speed of the Mexican wave is $v = \frac{d}{T} = 6\,\frac{\text{m}}{\text{s}}$, while the time interval is $\Delta t = \frac{T}{n} = 0.1\,\text{s}$.

(c) The maximal speed of the spectator is $v_{\max} = A\omega = 3.77\,\frac{\text{m}}{\text{s}}$, which is smaller than the speed of the Mexican wave v.

(d) As stated in the text of the problem, the pulse is a transverse wave that propagates along the x-axis without changing its shape. The height of the highest point of the spectator $y_p(x,t)$ depends on spatial coordinate x and time coordinate t. The shape of the pulse is defined by the function $f(x) = y_p(x,t = 0)$. Since the shape of the pulse is parabolic, it can be assumed that

$$f(x) = \begin{cases} h_1 & x \le -\frac{d_p}{2} \\ C_0 + C_1 x + C_2 x^2 & -\frac{d_p}{2} \le x \le \frac{d_p}{2} \\ h_1 & \frac{d_p}{2} \le x \end{cases}$$

where C_0, C_1, and C_2 are constants that should be determined. From Figure 6.2 we find that $f(-x) = f(x)$, $f(d_p/2) = h_1$, and $f_{\max} = f(0) = h_2$, which leads to $C_0 = h_2$, $C_1 = 0$, and $C_2 = \frac{4(h_1 - h_2)}{d_p^2}$.

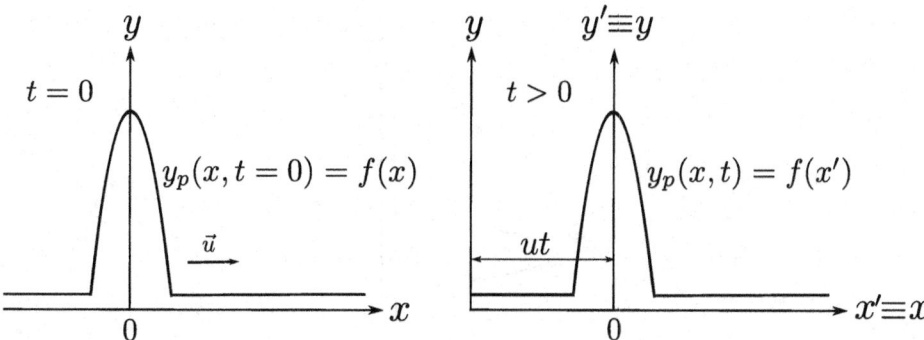

Figure 6.4 Propagation of the pulse of the Mexican wave

The center of the pulse will be at position $x = ut$ at time t. Since the shape of the pulse does not change, it follows that $y_p(x,t) = f(x - ut)$ (see Figure 6.4), and consequently

$$y_p(x,t) = \begin{cases} h_1 & x - ut \leq -\frac{d_p}{2} \\ h_2 - \frac{4(h_2 - h_1)}{d_p^2}(x - ut)^2 & -\frac{d_p}{2} \leq x - ut \leq \frac{d_p}{2} \\ h_1 & \frac{d_p}{2} \leq x - ut \end{cases}$$

(e) The time it takes the pulse to travel through n_p spectators is $T_p = \frac{d_p}{u}$, which implies $\Delta t = \frac{T_p}{n_p} = \frac{d_p}{u n_p} = 0.05$ s.

(f) The shape of the pulse at these times is shown in Figure 6.5.

(g) The speed of the highest point of the spectator is $v(x,t) = \left| \frac{dy_p(x,t)}{dt} \right|$ — that is,

$$v(x,t) = \begin{cases} 0 & t \leq \frac{x}{u} - \frac{d_p}{2u} \\ \frac{8u(h_2 - h_1)}{d_p^2} |x - ut| & \frac{x}{u} - \frac{d_p}{2u} \leq t \leq \frac{x}{u} + \frac{d_p}{2u} \\ 0 & \frac{x}{u} + \frac{d_p}{2u} \leq t \end{cases}$$

The graphs of the speed in question are shown in Figure 6.6.

(h) If the pulse was moving in the opposite direction, its equation would have the following form:

$$y_p(x,t) = \begin{cases} h_1 & x + ut \leq -\frac{d_p}{2} \\ h_2 - \frac{4(h_2 - h_1)}{d_p^2}(x + ut)^2 & -\frac{d_p}{2} \leq x + ut \leq \frac{d_p}{2} \\ h_1 & \frac{d_p}{2} \leq x + ut \end{cases}$$

(i) Since the wave front of the pulse is always in the direction of the radius vector $\vec{r}$, the angular velocities of the propagation of the pulse in the first and the last rows are equal and read $\omega_1 = \omega_2 = \dot{\phi}$. The speed of the body moving at an

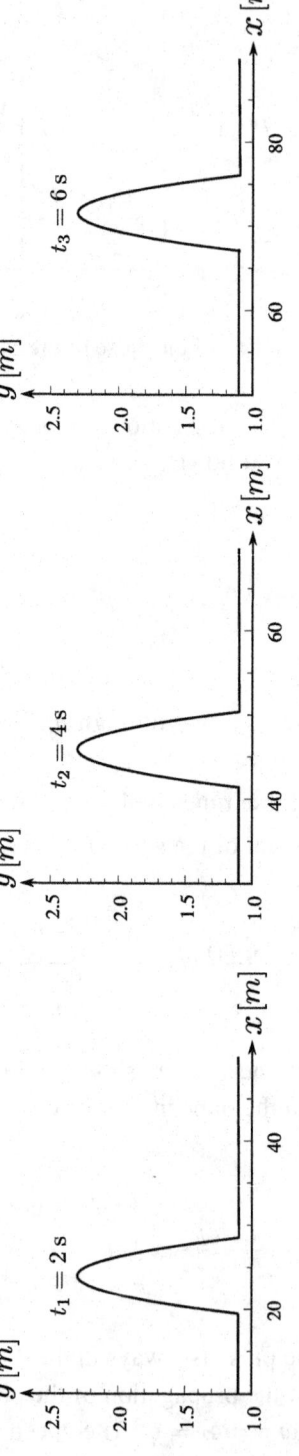

Figure 6.5   The wave pulse at times $t_1 = 2\,\text{s}$, $t_2 = 4\,\text{s}$, and $t_3 = 6\,\text{s}$, respectively

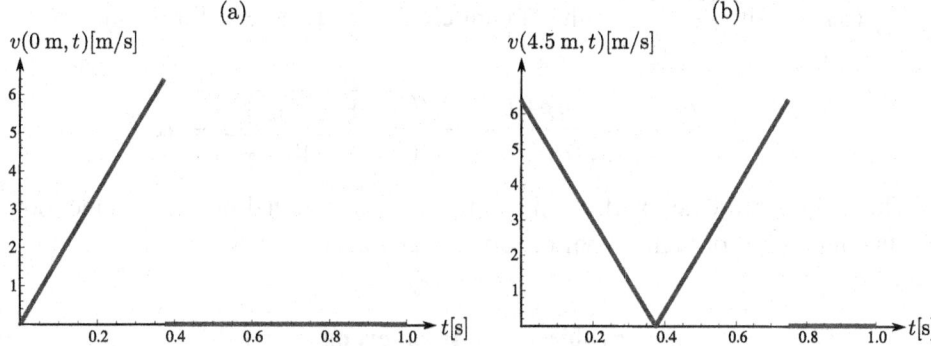

Figure 6.6   Time dependence of the speed of the highest point of the spectator at the position: (a) $x = 0$ m and (b) $x = 4.5$ m

elliptical trajectory $r(\varphi)$ reads $v(\varphi) = \sqrt{\dot{x}(\varphi)^2 + \dot{y}(\varphi)^2}$, where $x = r(\varphi)\cos\varphi$ and $y = r(\varphi)\sin\varphi$, which implies

$$
\begin{aligned}
v^2(\varphi) &= \left(\frac{dx}{d\varphi}\frac{d\varphi}{dt}\right)^2 + \left(\frac{dy}{d\varphi}\frac{d\varphi}{dt}\right)^2 \\
&= \left[\left(\frac{dr}{d\varphi}\cos\varphi - r\sin\varphi\right)^2 + \left(\frac{dr}{d\varphi}\sin\varphi + r\cos\varphi\right)^2\right]\dot{\varphi}^2 \\
&= \left[\left(\frac{dr}{d\varphi}\right)^2 + r^2\right]\dot{\varphi}^2 = \frac{a^2 b^2 (a^4 \sin^2\varphi + b^4 \cos^2\varphi)}{(a^2 \sin^2\varphi + b^2 \cos^2\varphi)^3}\dot{\varphi}^2 .
\end{aligned}
\tag{6.6}
$$

It follows from equation (6.6) that the speeds of the wave propagation in the first row u_1 and the last row $u_2(\varphi)$ satisfy

$$
u_1^2 = \frac{a_1^2 b_1^2 (a_1^4 \sin^2\varphi + b_1^4 \cos^2\varphi)}{(a_1^2 \sin^2\varphi + b_1^2 \cos^2\varphi)^3}\dot{\varphi}^2 ,
\tag{6.7}
$$

$$
u_2^2(\varphi) = \frac{a_2^2 b_2^2 (a_2^4 \sin^2\varphi + b_2^4 \cos^2\varphi)}{(a_2^2 \sin^2\varphi + b_2^2 \cos^2\varphi)^3}\dot{\varphi}^2 ,
\tag{6.8}
$$

which finally leads to

$$
u_2(\varphi) = \frac{a_2 b_2 (a_2^4 \sin^2\varphi + b_2^4 \cos^2\varphi)^{1/2}}{a_1 b_1 (a_1^4 \sin^2\varphi + b_1^4 \cos^2\varphi)^{1/2}} \frac{(a_1^2 \sin^2\varphi + b_1^2 \cos^2\varphi)^{3/2}}{(a_2^2 \sin^2\varphi + b_2^2 \cos^2\varphi)^{3/2}} u_1 .
\tag{6.9}
$$

(j) The time it takes for the Mexican wave to propagate through the whole stadium is $t = \frac{O_1}{u_1}$, where $O_1 \approx \pi[3(a_1 + b_1) - \sqrt{(3a_1 + b_1)(a_1 + 3b_1)}]$ is the circumference of the inner ellipse. The mean speed of the propagation of the pulse in the last row (outer ellipse) is $\bar{u}_2 = \frac{O_2}{t}$, where $O_2 \approx \pi[3(a_2 + b_2) -$

$\sqrt{(3a_2+b_2)(a_2+3b_2)}]$ is the circumference of the outer ellipse. One finally obtains

$$\bar{u}_2 = \frac{O_2}{O_1}u_1 \approx \frac{3(a_2+b_2)-\sqrt{(3a_2+b_2)(a_2+3b_2)}}{3(a_1+b_1)-\sqrt{(3a_1+b_1)(a_1+3b_1)}}u_1 \, .$$

The authors presented part of this problem at the national physics competition for the third grade of high school in Serbia in 2016.

Problem 39   Boomerang

What makes a boomerang go along a curved path?
What should be the ratio of its velocity and angular velocity so that it goes along a circular trajectory?

A boomerang is an object that has a curved path in the air. There is even a possibility that it returns to the person who threw it if it is thrown in an appropriate way. In this problem we consider a boomerang of mass m that consists of three arms, each of length l and width $b \ll l$, shown in Figure 6.7. The angle between two neighboring arms of the boomerang is $120°$. The boomerang was thrown in such a way that its center of mass travels on a circular trajectory in the horizontal plane; see Figure 6.8. In each moment of time the plane of the boomerang is vertical and the projection of the boomerang to the horizontal plane is tangential to the circular trajectory. Assume that the magnitude of the aerodynamic force acting on a small part of the boomerang of area ΔS is $\Delta F = \frac{1}{2}C_L \rho v_t^2 \Delta S$, where C_L is a dimensionless coefficient, ρ is the density of air, and v_t is the tangential component of the velocity of that part of the boomerang (more precisely, v_t is the projection of the velocity of that part on the line that is in the plane of the boomerang and is perpendicular to the line that connects the center of the boomerang and that part; see Figure 6.9). Aerodynamic force is directed perpendicularly to the plane of the boomerang. Neglect the gravity force and air resistance.

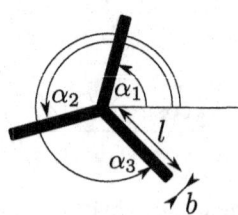

Figure 6.7 A boomerang that consists of three arms of length l and width b. The angle between the arm i and the horizontal is α_i

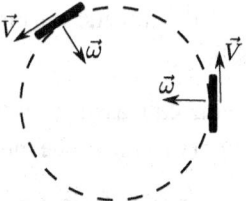

Figure 6.8 Top view of the trajectory of motion of the boomerang. The position of the boomerang, its velocity, and its angular velocity in two moments of time are shown in the figure

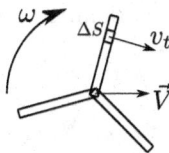

Figure 6.9 A small part of the boomerang of area ΔS and its tangential component of the velocity v_t

(a) Determine the moment of inertia of the boomerang with respect to the axis o that passes through its center and is perpendicular to the plane of the boomerang.

The boomerang was thrown in such a way that the speed of its center of mass is V, while the magnitude of the angular velocity of rotation around the axis o is ω, which is much larger than the magnitude of angular velocity of rotation of the center of mass around the center of the circular trajectory.

(b) Determine the magnitude of the total force that acts on the arm i of the boomerang in the moment when its angle with respect to horizontal is α_i.

(c) Determine the magnitude of the total force that acts on the boomerang at an arbitrary moment of time.

(d) Determine the torque (with respect to the center of the boomerang) that acts on the arm of the boomerang i in the moment when its angle with respect to horizontal is α_i.

(e) Determine the total torque (with respect to the center of the boomerang) that acts on the boomerang in the moment when arm 1 is at an angle of α_1 with respect to horizontal.

(f) Determine the mean value of this torque during one period of rotation of the boomerang around the axis o.

In the rest of the problem you can assume that the torque that acts on the boomerang is equal in each moment of time to the mean value determined in part (f).

(g) Determine the radius of the circular trajectory of the boomerang. Calculate the corresponding numerical value if $m = 40.0$ g, $l = 20.0$ cm, $b = 4.00$ cm, $C_L = 0.500$, and $\rho = 1.20 \frac{\text{kg}}{\text{m}^3}$.

(h) Determine the ratio of the quantities V and $l\omega$. The result should be in the form of a mathematical constant.

(i) Make an order-of-magnitude estimate of the radius of the circular trajectory of the boomerang without using the detailed analysis from previous parts of the problem.

Solution of Problem 39

(a) The moment of inertia of the arm i (the mass of which is m_i) with respect to the axis o is $I_i = \frac{1}{3}m_i l^2$, as follows from the formula for the moment of inertia of a homogeneous rod with respect to the axis that passes through the end of the rod and is perpendicular to the rod. The total moment of inertia of the boomerang is $I = I_1 + I_2 + I_3$, which leads to

$$I = \frac{1}{3}ml^2. \tag{6.10}$$

(b) The magnitude of the force that acts on the small part of the arm of length dr that is at a distance r from the center of the boomerang (Figure 6.10) is

$$dF = \frac{1}{2}C_L \rho b v_t^2 \, dr, \tag{6.11}$$

where

$$v_t = r\omega + V \sin \alpha_i. \tag{6.12}$$

The magnitude of the total force that acts on the arm i of the boomerang is

$$F_i = \int_0^l dF = \frac{1}{2}C_L \rho b \int_0^l dr \, (r\omega + V \sin \alpha_i)^2. \tag{6.13}$$

After integration, we obtain

$$F_i = \frac{1}{2}C_L \rho b \left(\frac{1}{3}\omega^2 l^3 + V^2 l \sin^2 \alpha_i + V\omega l^2 \sin \alpha_i \right). \tag{6.14}$$

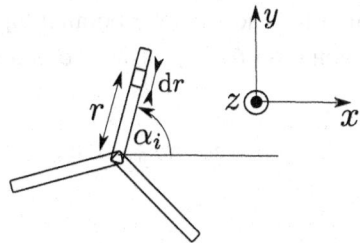

Figure 6.10 The boomerang, the part of the boomerang of length dr, and the coordinate system

(c) The magnitude of the total force that acts on the boomerang is $F = F_1 + F_2 + F_3$, where the forces F_i are given by equation (6.14). To calculate this sum we use the trigonometric identity

$$\sin^2 \alpha_i = \frac{1 - \cos 2\alpha_i}{2} \tag{6.15}$$

and the identities that are valid for three angles that are shifted by 120°

$$\sin \alpha_1 + \sin \alpha_2 + \sin \alpha_3 = 0, \tag{6.16}$$
$$\cos 2\alpha_1 + \cos 2\alpha_2 + \cos 2\alpha_3 = 0. \tag{6.17}$$

Using previous equations, we obtain

$$F = \frac{1}{4} C_L \rho l b \left(2\omega^2 l^2 + 3V^2 \right). \tag{6.18}$$

(d) The torque of the force that acts on the part of the arm of length dr that is at a distance r from the center of the boomerang (Figure 6.10) is

$$d\vec{M} = \vec{r} \times d\vec{F}, \tag{6.19}$$

where $\vec{r} = r(\cos \alpha_i \vec{e}_x + \sin \alpha_i \vec{e}_y)$ and $d\vec{F} = -dF\vec{e}_z$ (coordinate system is shown in Figure 6.10), where dF is given by equation (6.11). We then have

$$d\vec{M} = -dF \cdot r(\sin \alpha_i \vec{e}_x - \cos \alpha_i \vec{e}_y). \tag{6.20}$$

Since the torque of the force that acts on arm i is

$$\vec{M}_i = \int_0^l d\vec{M}, \tag{6.21}$$

we obtain after integration

$$\vec{M}_i = -(\sin \alpha_i \vec{e}_x - \cos \alpha_i \vec{e}_y) \frac{1}{2} C_L \rho b l^2 \left(\frac{1}{4} \omega^2 l^2 + \frac{1}{2} V^2 \sin^2 \alpha_i + \frac{2}{3} V \omega l \sin \alpha_i \right). \tag{6.22}$$

(e) Total torque of the force that acts on the boomerang is $\vec{M} = \vec{M}_1 + \vec{M}_2 + \vec{M}_3$, where M_i is given by equation (6.22). Using equations (6.15) and (6.16) and the analogous identity

$$\cos\alpha_1 + \cos\alpha_2 + \cos\alpha_3 = 0, \qquad (6.23)$$

it follows that

$$\vec{M} = -\tfrac{1}{2}C_L\rho b \sum_{i=1}^{3}\left[\tfrac{1}{4}V^2 l^2 \left(-\sin\alpha_i\cos 2\alpha_i \vec{e}_x + \cos\alpha_i\cos 2\alpha_i \vec{e}_y\right)\right.$$

$$\left. + \tfrac{2}{3}V\omega l^3 \left(\tfrac{1-\cos 2\alpha_i}{2}\vec{e}_x - \tfrac{\sin 2\alpha_i}{2}\vec{e}_y\right)\right]. \qquad (6.24)$$

We next use the identities

$$\sin\alpha\cos 2\alpha = \frac{1}{2}\left(\sin 3\alpha - \sin\alpha\right), \qquad (6.25)$$

$$\cos\alpha\cos 2\alpha = \frac{1}{2}\left(\cos 3\alpha + \cos\alpha\right), \qquad (6.26)$$

as well as equations (6.23), (6.16), and (6.17), and the analogous identity

$$\sin 2\alpha_1 + \sin 2\alpha_2 + \sin 2\alpha_3 = 0, \qquad (6.27)$$

which leads to

$$\vec{M} = -\frac{1}{2}C_L\rho b\left[\frac{1}{8}V^2 l^2\left(-\vec{e}_x\sum_{i=1}^{3}\sin 3\alpha_i + \vec{e}_y\sum_{i=1}^{3}\cos 3\alpha_i\right) + V\omega l^3 \vec{e}_x\right]. \qquad (6.28)$$

The last expression can be further simplified taking into account that $\cos 3\alpha_1 = \cos 3\alpha_2 = \cos 3\alpha_3$ and $\sin 3\alpha_1 = \sin 3\alpha_2 = \sin 3\alpha_3$, which leads to

$$\vec{M} = -\frac{1}{2}C_L\rho b\left[\frac{3}{8}V^2 l^2\left(-\vec{e}_x\sin 3\alpha_1 + \vec{e}_y\cos 3\alpha_1\right) + V\omega l^3 \vec{e}_x\right]. \qquad (6.29)$$

(f) The torque is given by equation (6.29) in each moment of time. The mean value of $\sin 3\alpha_1$ and $\cos 3\alpha_1$ during one period of rotation is zero. The mean torque is then

$$\langle \vec{M} \rangle = -\frac{1}{2}C_L\rho b V\omega l^3 \vec{e}_x. \qquad (6.30)$$

(g) We denote as ω_p the angular velocity of rotation of the center of mass around the center of the circular trajectory. This angular velocity is connected to the speed of the boomerang as

$$V = R\omega_p, \qquad (6.31)$$

where R is the radius of the circular orbit. Newton's second law for rotation gives $\vec{M} = \frac{d\vec{L}}{dt}$, where $\vec{L}$ is the angular momentum of the boomerang with respect to the center of mass. Since it was stated in the problem that $\omega_p \ll \omega$, the direction of the vector of angular momentum of the boomerang coincides in each moment of time with the direction of the vector $\vec{\omega}$. This direction connects the center of the boomerang and the center of the circular path. Consequently the vector $\vec{L}$ has constant magnitude, while its direction changes at an angular velocity ω_p. Therefore, we have

$$\left|\frac{d\vec{L}}{dt}\right| = \omega_p L, \tag{6.32}$$

which leads to

$$M = L\omega_p. \tag{6.33}$$

Using

$$L = I\omega, \tag{6.34}$$

it follows from equations (6.10), (6.31), (6.33), (6.30), and (6.34) that

$$R = \frac{2m}{3C_L \rho lb} = 5.56\,\text{m}. \tag{6.35}$$

(h) Newton's second law for the motion of boomerang on a circular trajectory gives

$$F = \frac{mV^2}{R}. \tag{6.36}$$

It follows from equations (6.18), (6.35), and (6.36) that

$$\frac{V}{l\omega} = \sqrt{\frac{2}{3}}. \tag{6.37}$$

(i) Aerodynamic force that acts on the small part of the boomerang is $\Delta F = \frac{1}{2}C_L \rho v_t^2 \Delta S$. The speed v_t has the same order of magnitude as the speed V. Therefore, the total force that acts on the boomerang can be estimated as $F \sim C_L \rho V^2 S$, where $S \sim bl$ is the surface area of the boomerang. In previous equations numerical factors of order 1 were not written because we are interested in an order-of-magnitude estimate only. Using Newton's second law $F = \frac{mV^2}{R}$, we obtain from previous equations $R \sim \frac{m}{C_L \rho lb}$. The result obtained agrees up to a numerical factor of order 1 with a result obtained from a detailed analysis.

We refer the reader interested in more details to reference [17]. This problem was created by modifying a problem given at a physics olympiad in Portugal in 2006.

Problem 40   Invisibility

How to make a material with negative electric permittivity.
How to make a material with negative magnetic permittivity.
How to use a material with negative refractive index to make an object invisible.

People have long dreamed of rendering objects invisible. One of the ways to make an object invisible is to shield it with such a cloak that light rays directed toward the object exhibit a curved path, bypass the object, and then return to their initial direction. A remote observer could not see the object in such a case and could not conclude that the object exists. Nevertheless, it is rather difficult to realize these invisibility cloaks in reality. It would be necessary to use a material with a complex spatial dependence of the electric and magnetic permittivity and it might be even necessary that these quantities and the refractive index take negative values.

We first investigate if negative electric permittivity is possible in natural materials.

We use a model of an atom where the nucleus is static. Its charge is q, while the electronic cloud is represented by a charged particle of charge $-q$ and mass m. An electronic cloud is connected to the nucleus by a spring of stiffness $m\omega_0^2$. The length of the spring is equal to zero when a force does not act on it. The drag force of magnitude $m\gamma v$ also acts on the particle that represents the electronic cloud. v is the speed of the particle, while $\gamma < \omega_0$. Assume that the electric field created due to the motion of electronic cloud does not act on other atoms of the material. The concentration of atoms is n.

(a) Determine the dependence of the relative electric permittivity of this material on angular frequency ω. Express the result in terms of the quantities $\omega_p^2 = \frac{nq^2}{m\varepsilon_0}$, ω_0, γ and ω. Help: Relative electric permittivity connects complex representatives of the polarization and the electric field by the relation $\underline{P}(\omega) = \varepsilon_0 \left[\varepsilon_r(\omega) - 1\right] \underline{E}(\omega)$. A complex representative of a physical quantity that depends on time as $x(t) = X_0 \cos(\omega t + \phi)$ is defined as $\underline{X}(\omega) = X_0 \exp(i\phi)$.

(b) Determine the condition that the quantities ω_p, ω_0, and γ should satisfy so that a range of frequencies exists where $\mathrm{Re}\,\varepsilon_r(\omega)$ is negative.

(c) In which range of angular frequencies is $\mathrm{Re}\,\varepsilon_r(\omega)$ negative if $\omega_p^2 = 0.1\,\omega_0^2$ and $\gamma = 0.01\,\omega_0$? Express the angular frequencies in question in units of ω_0.

The results of previous parts of the problem imply that it is very difficult to accomplish negative $\mathrm{Re}\,\varepsilon_r(\omega)$ in natural materials. On the other hand, it is possible to produce artificial materials and adjust their parameters to tailor the dependence $\varepsilon_r(\omega)$: these materials are called metamaterials.

We consider first a metamaterial that consists of an array of long metallic wires that form a square lattice, as shown in Figure 6.11. The distance between the axes of two neighboring wires is a, while the cross section of each wire is a circle of radius r, where $r \ll a$.

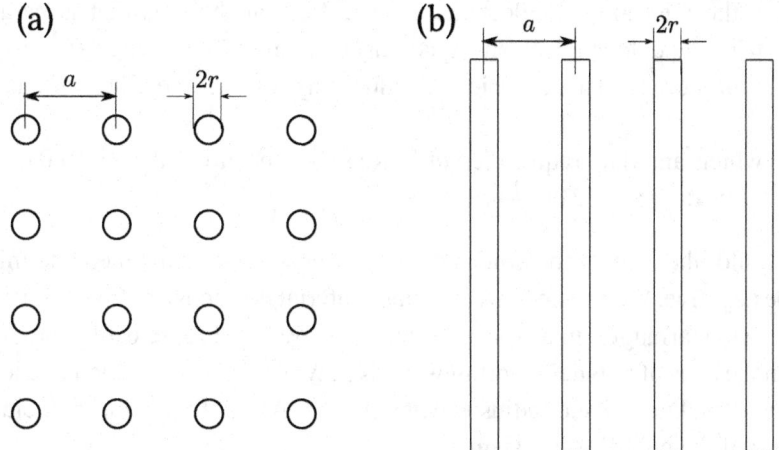

Figure 6.11 A scheme of the metamaterial considered in parts of the problem (d), (e), and (f). A top view is shown in part (a) of the figure, while a side view is shown in part (b) of the figure.

(d) Determine the flux of the magnetic field through a surface defined by the rectangular contour $ABCDA$ shown in Figure 6.12. The length of the rectangle

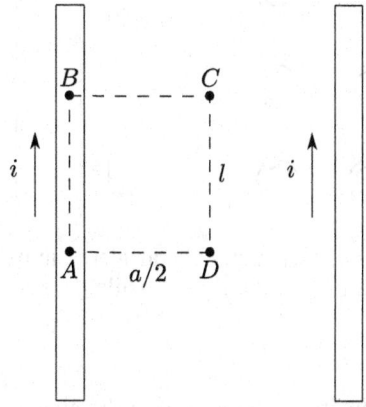

Figure 6.12 The surface defined by the contour $ABCDA$

is l, while its width is $a/2$. Assume that the current i passes through each of the wires and take into account only the magnetic field of two wires shown in Figure 6.12.

(e) When the system is excited by an electromagnetic field with a wavelength much longer than the distance between the wires, one can assume that the system of wires is a homogeneous metamaterial. Find $\varepsilon_r(\omega)$ of such a metamaterial. Assume that the electric field and polarization are directed along the wire axis. Neglect the electric field at the axis that is parallel to the wires and located at an equal distance from both wires (the axis CD in Figure 6.12). The electrical conductivity of the metallic wire material is σ.

(f) For which angular frequencies ω is $\operatorname{Re}\varepsilon_r(\omega)$ negative if $a = 10.0\,\text{mm}$, $r = 10.0\,\mu\text{m}$, $\sigma = 3.78\cdot 10^7\,\frac{1}{\Omega\cdot\text{m}}$?

We would like next to construct a material whose relative magnetic permittivity could be negative. We consider again a metamaterial that consists of an array of long metallic wires arranged in a square lattice, as shown in Figure 6.13. The distance between the axes of two neighboring wires is a, while the cross section of each wire is a ring of thickness b and radius r, where $b \ll r$. Axes of the wires coincide with the z-axis of the coordinate system.

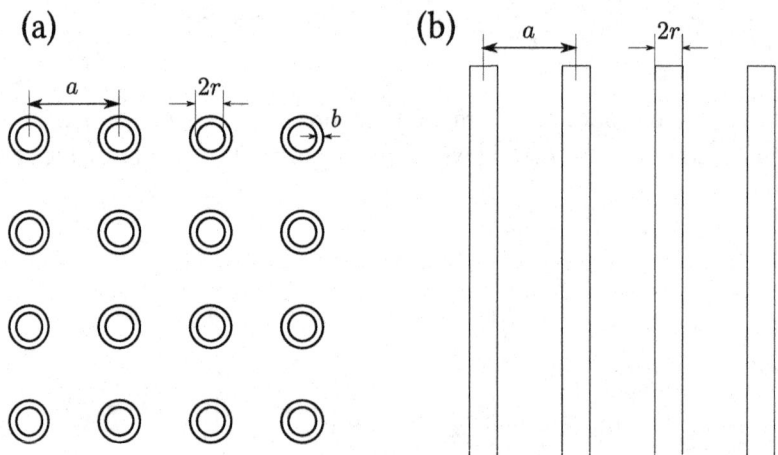

Figure 6.13 Scheme of the metamaterial considered in parts (g) through (j). The top view is shown in part (a) of the figure, while the side view in shown in part (b) of the figure

(g) The current of line density j is flowing on the surface of one wire in direction shown in Figure 6.14. Determine the magnetic field inside the wire caused by this current.

(a)   (b)

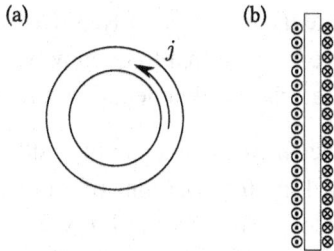

Figure 6.14 The direction of current flow through a wire considered in part (g) of the problem. The top view is shown in part (a) of the figure, while the side view in shown in part (b) of the figure

(h) Assume in this part of the problem that the current whose line density is j is flowing through all the wires except one wire, A. The current flows in the manner described in part (g) of the problem. Determine the z-component of magnetic field at the cross section of wire A. Assume that the z-component of the magnetic field in the plane of this cross section is homogeneous in the region outside the wires with current.

(i) Determine the relative magnetic permittivity $\mu_r(\omega)$ of the metamaterial that consists of this array of wires. Assume that the direction of the external magnetic field and the magnetization coincides with the wire axis. The electrical resistivity of the wire material is ρ.

(j) Show that $\mathrm{Re}\,\mu_r(\omega)$ obtained in part (i) is always positive.

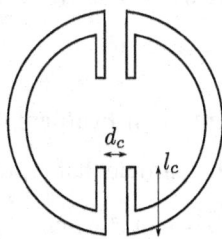

Figure 6.15 Cross section of the wires considered in parts (k) and (l)

(k) The wires considered in parts (g) through (j) were modified so that their cross section has the shape shown in Figure 6.15. The width of both of the two gaps is d_c, while the length of the two electrodes is l_c, where $l_c \gg d_c$. Determine the relative magnetic permittivity $\mu_r(\omega)$ of the metamaterial that consists of this array of wires.

(l) What is the range of angular frequencies where the condition $\mathrm{Re}\,\mu_r(\omega) < 0$ is satisfied if $r = 2.00\,\mathrm{mm}$, $a = 5.00\,\mathrm{mm}$, $d_c = 0.100\,\mathrm{mm}$, $l_c = 1.00\,\mathrm{mm}$, and $\rho/b = 0.100\,\Omega$?

The material with both $\mathrm{Re}\,\mu_r(\omega) < 0$ and $\mathrm{Re}\,\varepsilon_r(\omega) < 0$ at a certain frequency exhibits a negative refractive index at that frequency. We consider next some of the effects that can appear in materials with a negative refractive index.

(m) An object is at a distance a from the plate of width d made of a material with refractive index of -1. Find the positions of all images that are formed.

(n) To make a tank invisible soldiers covered it with a cloak whose cross section is the trapezoid $ABCD$ shown in Figure 6.16. The tank is at the middle of side AB. The angle between side AD and the vertical is $\alpha = 30.0°$. The plane δ is perpendicular to the line AB and divides it into two equal parts. The area on the left from the plane δ is filled with a material of refractive index $n = 1.30$, while the refractive index in the area on the right from that plane is $-n$. Determine the angle θ between the side BC and the vertical that makes the tank invisible for all observers on the ground (i.e., on the line AB).

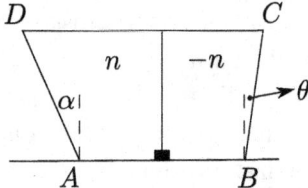

Figure 6.16 The cloak whose cross section is the trapezoid $ABCD$

The speed of light in a vacuum is $c = 3.00 \cdot 10^8\ \frac{\mathrm{m}}{\mathrm{s}}$, while the vacuum permittivity is $\varepsilon_0 = 8.85 \cdot 10^{-12}\ \frac{\mathrm{F}}{\mathrm{m}}$.

Solution of Problem 40

(a) The equation of motion of the particle that represents the electronic cloud is

$$m\ddot{\vec{r}} = -m\omega_0^2\vec{r} - m\gamma\dot{\vec{r}} - q\vec{E}(t),\tag{6.38}$$

where $\vec{E}(t) = E_0\cos(\omega t + \phi)\,\vec{e}_x$ is the external electric field. The dot and the double dot above the vector $\vec{r}$ denote respectively the first and the second derivative with respect to time t. It was assumed, without losing generality, that the electric field is in the x-direction and $\vec{e}_x$ is the unit vector in that direction. The projection of the equation of motion to the x-direction gives:

$$m\ddot{x} = -m\omega_0^2 x - m\gamma\dot{x} - qE_0\cos(\omega t + \phi).\tag{6.39}$$

The relation between complex representatives of the quantities $x(t)$ and $E_x(t)$ is then:

$$-m\omega^2\underline{x} = -m\omega_0^2\underline{x} - im\gamma\omega\underline{x} - q\underline{E}_x. \tag{6.40}$$

The x-component of the electric dipole moment of one atom is $p_x = -qx$. Therefore, the x-component of the polarization of the material is $P_x = -nqx$. Using equation (6.40) it follows that

$$\underline{P}_x = \frac{nq^2}{m} \frac{\underline{E}_x}{\omega_0^2 - \omega^2 + i\gamma\omega}. \tag{6.41}$$

The previous equation directly leads to

$$\varepsilon_r(\omega) = 1 + \frac{\omega_p^2}{\omega_0^2 - \omega^2 + i\gamma\omega}. \tag{6.42}$$

(b) It follows from equation (6.42) that

$$\operatorname{Re}\varepsilon_r(\omega) = 1 + \frac{\omega_p^2(\omega_0^2 - \omega^2)}{(\omega_0^2 - \omega^2)^2 + (\gamma\omega)^2}. \tag{6.43}$$

The condition $\operatorname{Re}\varepsilon_r(\omega) < 0$ yields

$$\omega^4 + (\gamma^2 - \omega_p^2 - 2\omega_0^2)\omega^2 + \omega_0^4 + \omega_p^2\omega_0^2 < 0. \tag{6.44}$$

The expression on the left-hand side of equation (6.44) is a quadratic polynomial with respect to ω^2. The inequality (6.44) will be satisfied for some ω if its discriminant is positive

$$(\gamma^2 - \omega_p^2 - 2\omega_0^2)^2 - 4(\omega_0^4 + \omega_p^2\omega_0^2) > 0. \tag{6.45}$$

From equation (6.45) it follows that

$$|\omega_p^2 - \gamma^2| > 2\gamma\omega_0. \tag{6.46}$$

(c) $\operatorname{Re}\varepsilon_r(\omega)$ is negative in the range of angular frequencies $\omega \in (\omega_1, \omega_2)$ where ω_1 and ω_2 are respectively the smaller and larger solution of the quadratic equation

$$\omega^4 + (\gamma^2 - \omega_p^2 - 2\omega_0^2)\omega^2 + \omega_0^4 + \omega_p^2\omega_0^2 = 0. \tag{6.47}$$

It then follows

$$\omega_{1,2}^2 = \frac{2\omega_0^2 + \omega_p^2 - \gamma^2 \mp \sqrt{(\gamma^2 - \omega_p^2)^2 - 4\gamma^2\omega_0^2}}{2}. \tag{6.48}$$

After the replacement of the values for γ and ω_p given in the text of the problem, we find $\omega \in (1.00051, 1.048)\,\omega_0$.

(d) We first choose the coordinate system as shown in Figure 6.17. The magnetic field in point P has only the z-component, which is equal to $B = -\left[\frac{\mu_0 i}{2\pi x} - \frac{\mu_0 i}{2\pi(a-x)}\right]$. The first term originates from the left wire, while the second originates from the right wire. The flux through the small area $l \cdot dx$ shown in Figure 6.17 is $d\Phi = -B \cdot l \cdot dx$. The total flux through the area bounded by the contour is obtained by performing the integration $\Phi = -\int_r^{a/2} B(x) l \cdot dx$. This leads to $\Phi = \frac{\mu_0 \cdot i \cdot l}{2\pi} \int_r^{a/2} dx \left(\frac{1}{x} - \frac{1}{a-x}\right)$ – that is, $\Phi = \frac{\mu_0 \cdot i \cdot l}{2\pi}\left(\ln\frac{a}{2r} - \ln\frac{a-r}{a/2}\right)$. Using $r \ll a$ it follows that

$$\Phi = \frac{\mu_0 \cdot i \cdot l}{2\pi} \ln\frac{a}{4r}. \tag{6.49}$$

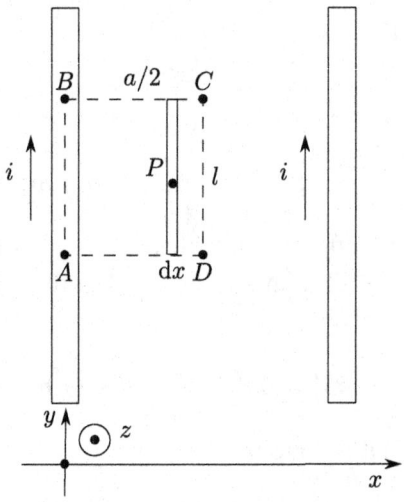

Figure 6.17 The view of the area through which the flux is calculated

(e) The current that flows through the wire is $i = r^2\pi nqv_y$, where n is the concentration of carriers in the metal, q is the charge of one carrier, and v_y is their average drift velocity. The y-component of the polarization of the metamaterial is $P_y = \frac{nr^2\pi}{a^2}qy$, where $\frac{nr^2\pi}{a^2}$ is the concentration of the carriers in the metamaterial and y is the mean displacement of a carrier from its equilibrium position. Using $v_y = \frac{dy}{dt}$ and the last two equations, we find

$$i = \frac{dP_y}{dt}a^2. \tag{6.50}$$

The induced electromotive force in the contour $ABCDA$ is $\varepsilon_i = -\frac{d\Phi}{dt}$. It then follows from equation (6.49) that

$$\varepsilon_i = -\frac{\mu_0 \cdot l}{2\pi}\frac{di}{dt}\ln\frac{a}{4r}. \qquad (6.51)$$

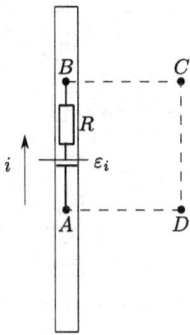

Figure 6.18 Equivalent electrical scheme of the part AB of the contour

The induced electromotive force in the contour $ABCDA$ is equal to the sum of the induced electromotive forces in each of the parts of the contour $\varepsilon_i = \varepsilon_i\,(AB) + \varepsilon_i\,(BC) + \varepsilon_i\,(CD) + \varepsilon_i\,(DA)$. It was stated in the problem that the electric field on the line CD can be neglected, hence $\varepsilon_i\,(CD) = 0$. Symmetry of the problem leads to $\varepsilon_i\,(BC) = -\varepsilon_i\,(DA)$ and consequently $\varepsilon_i = \varepsilon_i\,(AB)$. The y-component of the electric field in the wire is $E_y = \frac{V_A - V_B}{l}$, where (see Figure 6.18) $V_A - V_B = -\varepsilon_i\,(AB) + iR$ and $R = \frac{l}{r^2\pi\sigma}$ is the resistance of the part of the wire AB. We then obtain

$$E_y = -\frac{\varepsilon_i}{l} + \frac{i}{r^2\pi\sigma}. \qquad (6.52)$$

By replacing the expression for i from equation (6.50) to equations (6.51) and (6.52), and the expression for ε_i from equation (6.51) to equation (6.52) we obtain

$$\frac{\mu_0 a^2}{2\pi}\ln\frac{a}{4r}\ddot{P}_y + \frac{a^2}{r^2\pi\sigma}\dot{P}_y - E_y = 0. \qquad (6.53)$$

The relation between the complex representative of the quantities P_y and E_y is then

$$-\omega^2\frac{\mu_0 a^2}{2\pi}\ln\frac{a}{4r}\underline{P}_y\,(\omega) + i\omega\frac{a^2}{r^2\pi\sigma}\underline{P}_y\,(\omega) - \underline{E}_y\,(\omega) = 0, \qquad (6.54)$$

and the relative electric permittivity is

$$\varepsilon_r(\omega) = 1 - \frac{1}{\frac{\omega^2 a^2}{2\pi c^2}\ln\frac{a}{4r} - i\omega\varepsilon_0\frac{a^2}{\sigma r^2 \pi}}. \tag{6.55}$$

(f) It follows from equation (6.55) that

$$\mathrm{Re}\,\varepsilon_r(\omega) = 1 - \frac{\frac{\omega^2 a^2}{2\pi c^2}\ln\frac{a}{4r}}{\left(\frac{\omega^2 a^2}{2\pi c^2}\ln\frac{a}{4r}\right)^2 + \left(\omega\varepsilon_0\frac{a^2}{\sigma r^2 \pi}\right)^2}. \tag{6.56}$$

From the condition $\mathrm{Re}\,\varepsilon_r(\omega) < 0$ we find

$$\omega < \sqrt{\frac{\frac{a^2}{2\pi c^2}\ln\frac{a}{4r} - \left(\frac{a^2\varepsilon_0}{\sigma r^2\pi}\right)^2}{\left(\frac{a^2}{2\pi c^2}\ln\frac{a}{4r}\right)^2}}, \tag{6.57}$$

which leads to $\omega < 32.0\,\mathrm{GHz}$.

(g) Since the current flows on the surface of the wire and its direction is axial, the magnetic field inside the wire is the same as the magnetic field of a long coil. This field is directed along the z-axis and its z-component is $B = \mu_0 j$.

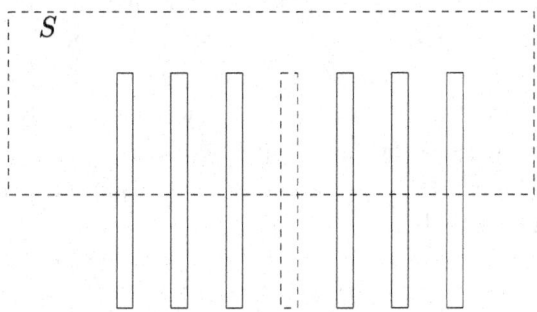

Figure 6.19 Closed surface S for application of the magnetic flux conservation law in part (h) of the problem

(h) We apply the law of conservation of magnetic flux to closed surface S shown in Figure 6.19. This surface was chosen so that the plane of its bottom base passes through all wires and is perpendicular to the wire axes, while the plane of the upper base and side surfaces are sufficiently far from the wires so that the magnetic field on these parts of the surface can be neglected. The flux through the surface S is then $\Phi \approx -Br^2\pi(N-1) + B_z N a^2$, where B was found in part (g), N is the number of wires, and B_z is the component of the magnetic field in question. In accordance with the text of the problem, it was assumed that

B_z is homogeneous. From $\Phi = 0$, using $N - 1 \approx N$ for large N, we find $B_z = -\mu_0 j \frac{r^2\pi}{a^2}$.

(i) To determine the relative magnetic permittivity of the metamaterial, we apply the magnetic field $B_0(t)$ along the z-axis. The z-component of the magnetic field inside the wire is $B_z = B_0 + \mu_0 j - \mu_0 j \frac{r^2\pi}{a^2}$, where the first term originates from an external field, the second from the field of the wire, and the third from the field of other wires (as shown in the previous part of the problem). The induced electromotive force in the closed contour defined by the cross section of the wire is then $\varepsilon_i = -\frac{dB_z}{dt} r^2\pi$. The drop of voltage in that contour is on the other hand equal to $U_R = (j \cdot dl) \cdot \frac{\rho \cdot 2r\pi}{b \cdot dl}$, where $j \cdot dl$ is the current through the contour, $\frac{\rho \cdot 2r\pi}{b \cdot dl}$ is the resistance of the contour, and dl is its length in the z-direction. Since $U_R = \varepsilon_i$, it follows that:

$$- i\omega r^2 \pi \left[\underline{B_0}(\omega) + \mu_0 \underline{j}(\omega) - \mu_0 \underline{j}(\omega) \frac{r^2\pi}{a^2} \right] = \underline{j}(\omega) \frac{\rho}{b} 2r\pi. \tag{6.58}$$

The magnetic moment of the part of a wire of length dl is $dm = j \cdot dl \cdot r^2\pi$. It is only this part of the wire that contributes to magnetization of the metamaterial in the part of space in the shape of a cuboid whose base is a square of side a, and the height is dl. The z-component of the magnetization of the metamaterial is therefore $M_z = \frac{dm}{dV}$, where $dV = a^2 dl$. This leads to $M_z = \frac{jr^2\pi}{a^2}$ and

$$\underline{M_z}(\omega) = \frac{\underline{j}(\omega) r^2\pi}{a^2}. \tag{6.59}$$

From the definition of relative magnetic permittivity it follows that

$$\underline{M_z}(\omega) = \frac{B_0}{\mu_0} \left(1 - \frac{1}{\mu_r(\omega)} \right). \tag{6.60}$$

From equations (6.58), (6.59), and (6.60) it follows that

$$\mu_r(\omega) = 1 - \frac{r^2\pi/a^2}{1 - i\frac{2\rho}{b\omega\mu_0 r}}. \tag{6.61}$$

(j) From equation (6.61) it follows $\mathrm{Re}\,\mu_r(\omega) = 1 - \frac{r^2\pi/a^2}{1 + \left(\frac{2\rho}{\omega\mu_0 r}\right)^2}$. Since the second term is always smaller than 1, it follows that $\mathrm{Re}\,\mu_r(\omega) > 0$.

(k) The two gaps introduced in the wire can be considered as two capacitors connected in a series. The capacitance of both of them is $C = \varepsilon_0 \frac{l_c \cdot dl}{d_c}$. The complex representative of the total voltage drop on these two capacitors is $\underline{U_C}(\omega) = \frac{-i}{C_{ekv}\omega} \cdot \underline{j}(\omega) dl$, where $C_{ekv} = \frac{1}{2}C$ is the equivalent capacitance of two capacitors connected in series. This leads to $\underline{U}_C(\omega) = \frac{-i\cdot 2d_c}{\varepsilon_0 l_c \omega}\underline{j}(\omega)$. Since $\varepsilon_i = U_C + U_R$, it follows that

$$-i\omega r^2 \pi \left[\underline{B}_0(\omega) + \mu_0 \underline{j}(\omega) - \mu_0 \underline{j}(\omega)\frac{r^2\pi}{a^2}\right] = \underline{j}(\omega)\frac{\rho}{b}2r\pi + \frac{-i\cdot 2d_c}{\varepsilon_0 l_c \omega}\underline{j}(\omega).$$

(6.62)

From equations (6.62), (6.59) and (6.60) we obtain

$$\mu_r(\omega) = 1 - \frac{f\omega^2}{\omega^2 - \omega_0^2 - i\Gamma\omega},$$

(6.63)

where $f = \frac{r^2\pi}{a^2}$, $\Gamma = \frac{2\rho}{b\mu_0 r}$, $\omega_0^2 = \frac{2d_c\cdot c^2}{r^2\pi l_c}$.

(l) From equation (6.63) it follows that $\mathrm{Re}\,\mu_r(\omega) = 1 - \frac{f\omega^2(\omega^2-\omega_0^2)}{(\omega^2-\omega_0^2)^2+\Gamma^2\omega^2} < 0$. After elementary transformations, we obtain that this inequality is equivalent to $\omega^4(1-f) + \omega^2(\Gamma^2 - 2\omega_0^2 + \omega_0^2 f) + \omega_0^4 < 0$. The last inequality is satisfied for $\omega \in (\omega_1, \omega_2)$, where $\omega_{1,2}^2 = \frac{\omega_0^2(2-f)-\Gamma^2\mp\sqrt{\omega_0^4 f^2+\Gamma^4-2\Gamma^2\omega_0^2(2-f)}}{2(1-f)}$. We finally obtain $\omega \in (37.8, 53.7)$ GHz.

(m) The paths of two characteristic rays emitted by the object in the case $a < d$ are shown in Figure 6.20(a). Since the refractive index of the plate is -1, the angle between the lines AF and p is equal to the angle between the lines CF

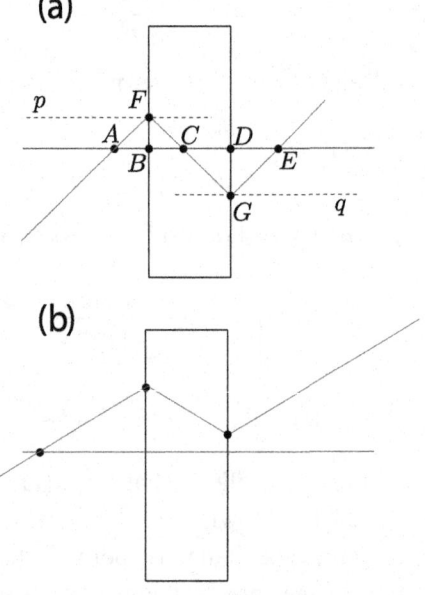

Figure 6.20 The paths of two characteristic rays when (a) $a < d$; (b) $a > d$

and p. For the same reason, the angle between CG and q is equal to the angle between GE and q. The triangles AFC and CGE are therefore isosceles and it holds that $AB = BC$ and $CD = DE$. The images are formed at points C and E, while $BC = AB = a$ and $DE = CD = d - BC = d - a$. In the case $a = d$ the points C and E are the same and only one image is formed. The paths of two characteristic rays in the case $a > d$ are shown in Figure 6.20(b). It can be seen from the figure that the paths do not cross and consequently no image is formed.

(n) The incident ray first exhibits a refraction at AD, then it refracts at the plane δ and finally refracts at BC, as shown in Figure 6.21. From Snell's law for refraction at AD we find $\sin \alpha = n \sin \beta$, where β is the angle between p and AE (where p is perpendicular to AD, and E is the point where the ray passes through the plane δ). The incident angle of the ray is $\alpha - \beta$ when it refracts at the plane δ. The angle between the refracted ray (whose path is the line EB) and the horizontal is the same because the refractive index changes only its sign at the plane δ. We define line q to be perpendicular to BC and we denote the angle between BE and q as ϕ. From Snell's law at BC we find $n \sin \phi = \sin \gamma$, where γ is the angle between the refracted ray and the line q. The tank is invisible if the direction of the refracted ray is horizontal – that is, if the angle γ is equal to the acute angle $\angle qBA$, which leads to $\gamma = \alpha - \beta - \phi$. Previous equations lead to $n \sin \phi = \sin (\alpha - \beta - \phi)$. The numerical solution of this equation gives $\phi = 3.21°$. The angle between BC and the vertical is then $\theta = \alpha - \beta - \phi = 4.17°$. Consequently the tank will be invisible for all observers on the ground for this θ. Nevertheless, to ensure the invisibility of the tank for an arbitrary observer in space, a cloak significantly more complicated than this one needs to be made.

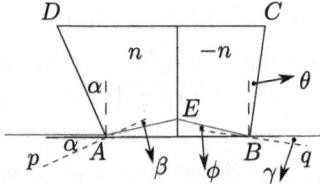

Figure 6.21 Refraction of rays at the invisibility cloak

We refer the interested reader to reference [27].

Problem 41   Soap Bubbles

Why is a soap bubble spherical?
How does the bubble deform when merged with other bubbles?

One of children's favorite activities is making soap bubbles by blowing air into a thin soap film. We investigate the physics behind this phenomenon in this problem. The surface tension of soap is γ, while the density of soap is ρ_s. The width of the soap layer in the bubble is much smaller than the radius of the bubble. The atmospheric pressure of air is p_0, the temperature is T_0, and the density of air is ρ_0. Gravitational acceleration is g.

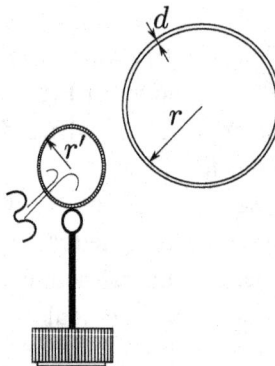

Figure 6.22 Soap bubble of radius r and width d made by blowing into the soap layer in the ring of radius r'

(a) A soap bubble of radius r was made by blowing into the soap layer inside the ring of radius r', shown in Figure 6.22. Determine the work A against capillary forces that needs to be performed to make such a bubble.

(b) What is the air pressure inside the soap bubble of radius r?

(c) A soap bubble that levitates in air can be made by blowing warm air into the soap film. What is the temperature of the air in the bubble in that case? The radius of levitating bubble is r, while its width is d.

(d) We know from experience that a soap bubble will not form in each case when we blow into the soap layer in the ring. It turns out that the bubble will form when the pressure in the soap bubble is equal to the pressure that would act on the wall that stops the air that flows in the direction perpendicular to the wall. Determine the minimal speed v of air flowing into the soap layer in the ring that leads to the formation of soap bubble of radius r. Assume that the profile of the velocity of flowing air is constant – that is, that each small part of air along the same cross section has the same velocity.

(e) It sometimes happens that two bubbles merge into one. It can be assumed that this process is isothermal. Show that the volume of the new bubble is equal to the sum of the volumes of the two bubbles when outside pressure is atmospheric p_0, while the surface tension of soap is very small ($\gamma \to 0$). Next show that the surface area of the new bubble is equal to the sum of the surface areas of the two bubbles in the case when atmospheric pressure is very low ($p_0 \to 0$), while the surface tension is finite.

(f) If two identical bubbles were connected using a straw of negligible volume, the system would be in equilibrium. Is this equilibrium stable?

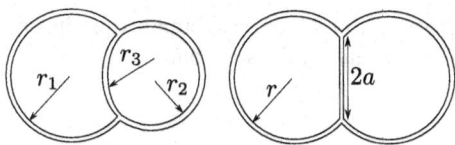

Figure 6.23 (left) Two connected bubbles of different radii. (right) Two connected bubbles of equal radii

(g) Two soap bubbles can be merged into two bubbles with a common surface, as shown in Figure 6.23. The radii of the smaller and the larger bubbles are, respectively, r_1 and r_2. What is the radius of curvature of the common surface r_3? How does this result change if $r_1 = r_2$?

(h) Determine the angles between three soap layers at their joint.

(i) A soap bubble is spherical because the surface energy of soap is smallest for this shape out of all other possible shapes with a given volume of air. In a similar manner, two bubbles merge into two bubbles with a common surface because they reduce the surface energy this way. The common surface is a disk of radius a, while the rest of the bubbles is spherical with radius r (Figure 6.23). Determine the ratio a/r in realistic conditions – that is, when the surface energy of the soap is minimal for the given volume of the system V.

Solution of Problem 41

(a) The work against capillary forces is proportional to the change of free surface of soap film $A = \gamma(S_2 - S_1)$, where $S_1 = 2\pi r'^2$ and $S_2 = 2 \cdot 4\pi r^2$ (factor 2 originates from the fact that the film has two free surfaces of approximately equal area since $d \ll r$), which leads to $A = 2\pi\gamma(4r^2 - r'^2)$.

(b) We consider one half of the spherical soap bubble of radius r shown in Figure 6.24. The force F_p that originates from the difference of pressures inside and outside the bubble given as $F_p = (p - p_0)\pi r^2$ acts on the half sphere vertically upward. The capillary force $F_\gamma = 4\gamma\pi r$ acts on the half sphere vertically downward. From the equilibrium condition $F_p = F_\gamma$, we find that the pressure inside the soap bubble is $p = p_0 + \frac{4\gamma}{r}$.

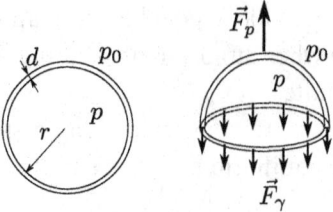

Figure 6.24 (left) Cross section of the soap bubble of radius r and width d. (right) The forces acting on upper half of the soap bubble

(c) The bubble floats when its weight is balanced by the buoyancy force $(m_s + m_v)g = \rho_0 \frac{4}{3}\pi r^3 g$. Since $d \ll r$, the mass of the soap is $m_s = \rho_s 4\pi r^2 d$. The mass of air in the bubble can be calculated from the ideal gas law $\left(p_0 + \frac{4\gamma}{r}\right)\frac{4}{3}\pi r^3 = \frac{m_v}{M}RT$, where R is the gas constant, while M is the molar mass of air that can be calculated from the ideal gas law of the surrounding air in the atmosphere $p_0 = \frac{R\rho_0 T_0}{M}$. We obtain from previous equations that the temperature of air in the bubble is $T = \frac{p_0 T_0 (r p_0 + 4\gamma)}{p_0(\rho_0 r - 3\rho_s d)}$.

(d) To calculate the pressure of air on the wall of area S, we consider the motion of a beam of particles of speed v directed perpendicularly to the wall. All particles from the volume $dV = Svdt$ hit the wall during time dt and transfer the momentum $dP = vdm = \rho_v v^2 Sdt$ to the wall. The pressure on the wall is therefore $p_1 = p_0 + \rho_v v^2$. This pressure is equal to the pressure inside the formed bubble $p_2 = p_0 + \frac{4\gamma}{r}$, which leads to the expression for minimal blowing speed $v = \sqrt{\frac{4\gamma}{\rho_v r}}$.

(e) We denote as r_1 and r_2 the radii of initial soap bubbles, while r_3 is the radius of new bubble. The ideal gas law for each of the bubbles reads $p_i V_i = \left(p_0 + \frac{4\gamma}{r_i}\right)\frac{4}{3}\pi r_i^3 = n_i R T_0$, for $i = \{1,2,3\}$. Since the total amount of gas does not change – that is, $n_1 + n_2 = n_3$ – we have

$$\left(p_0 + \frac{4\gamma}{r_1}\right)r_1^3 + \left(p_0 + \frac{4\gamma}{r_2}\right)r_2^3 = \left(p_0 + \frac{4\gamma}{r_3}\right)r_3^3. \qquad (6.64)$$

When the atmospheric pressure is p_0 – that is, it is finite, while surface tension of the soap is small ($\gamma \to 0$), equation (6.64) becomes $r_1^3 + r_2^3 = r_3^3$, which leads to $V_1 + V_2 = V_3$. When the atmospheric pressure is very low ($p_0 \to 0$), while the surface tension of the soap is finite, equation (6.64) becomes $r_1^2 + r_2^2 = r_3^2$, which leads to $S_1 + S_2 = S_3$.

(f) If the radius of one bubble decreases and in the other it increases, the pressure in the first, smaller bubble increases while the pressure in the second, larger bubble decreases. This difference of pressures causes the smaller bubble to further decrease its size and the larger bubble to further increase its size. This continues until the smaller bubble completely transforms into the larger bubble. Consequently the equilibrium is not stable.

(g) The pressure inside the larger bubble is $p_1 = p_0 + \frac{4\gamma}{r_1}$, while the pressure inside the smaller bubble is $p_2 = p_0 + \frac{4\gamma}{r_2}$. The difference of these pressures is $p_2 - p_1 = \frac{4\gamma}{r_3}$, which leads to $r_3 = \frac{r_1 r_2}{r_1 - r_2}$. If $r_1 = r_2$, the radius of curvature of the common surface is $r_3 \to \infty$ – that is, the surface is flat. Next it follows from symmetry considerations that it is in the shape of a circle.

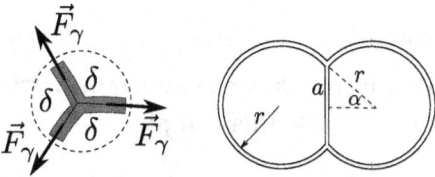

Figure 6.25 (left) Equilibrium of the part of the soap at the joint of three soap layers. (right) Two merged bubbles of equal radii

(h) We consider the equilibrium of three small parts of soap at the joint of three soap layers, as shown in Figure 6.25. The capillary force of magnitude $F_\gamma = 2\gamma \Delta l$ acts on each part of length Δl. Each of these forces is tangential to the corresponding layer. Since the magnitude of these forces is the same, it follows that the angles between two neighboring layers are $\delta = 2\pi/3$.

(i) Since $a/r = \sin\alpha$ (Figure 6.25) we describe the system using the parameters r and α. The total volume of the system is $V = 2(V' - V'')$, where $V' = \frac{4}{3}\pi r^3$ is the volume of the sphere, while $V'' = \frac{1}{3}\pi h^2(3r - h)$ is the volume of the calotte of height $h = r(1 - \cos\alpha)$. The connection between r and α when the volume of the system is constant is

$$r^3 = \frac{3V}{2\pi} \frac{1}{4 - 3(1 - \cos\alpha)^2 + (1 - \cos\alpha)^3}. \tag{6.65}$$

The surface energy of the soap is $E = 4\gamma S' + 2\gamma S''$, where $S' = 4\pi r^2 - 2\pi r h$ is the surface of the part of the sphere without the calotte of height h, while $S'' = \pi a^2$ is the common surface of two bubbles. Using equation (6.65) we obtain the energy

$$E(\alpha) = (2\pi)^{\frac{1}{3}}(3V)^{\frac{2}{3}}\gamma \frac{4 + 4\cos\alpha + \sin^2\alpha}{(4 - 3(1 - \cos\alpha)^2 + (1 - \cos\alpha)^3)^{\frac{2}{3}}}. \tag{6.66}$$

The first derivative of this function is

$$E'(\alpha) = (2\pi)^{\frac{1}{3}}(3V)^{\frac{2}{3}}\gamma\frac{2\sin\alpha(2\cos\alpha-1)}{(-2+\cos\alpha)(2+3\cos\alpha-\cos^3\alpha)^{\frac{2}{3}}}, \qquad (6.67)$$

which leads to $\partial E/\partial\alpha = 0$ for $\alpha \in \{0, \pi/3\}$. Since $\partial^2 E/\partial\alpha^2 > 0$ only for $\alpha = \pi/3$, we conclude that this is the value when $E(\alpha)$ reaches minimum. The ratio in question is therefore $a/r = \sin\alpha = \sqrt{3}/2$.

We refer interested readers to references [6] and [28].

Problem 42   Champagne Bottle Opening

Why do small clouds appear when a champagne bottle is opened? What height does a cork reach after popping out?

Champagne is a popular drink consumed on occasions when people celebrate an important event, such as completing a collection of *Fascinating Problems for Young Physicists*. When a champagne bottle is opened, spectacular popping out of the cork is accompanied by the appearance of small clouds of condensed water vapor near the top of the bottle throat. Unlike other wines, champagne is bottled before the completion of fermentation. As a result, fermentation in the bottle leads to the appearance of carbon dioxide gas at a high pressure in the region of volume $V_1 = 5.00\,\mathrm{cm}^3$ below the cork and above the champagne. We assume that the throat of the bottle is cylindrical. Its inner diameter is $d = 16.2\,\mathrm{mm}$, while the density of the cork is $\rho = 995\,\frac{\mathrm{kg}}{\mathrm{m}^3}$. The cork is in the shape of a mushroom; see Figure 6.26. The bottom part of the cork that is in the bottle is of a cylindrical shape with diameter d and height $h = d$. The top part of the cork that is outside the bottle is of a hemispherical shape with radius $r = 3d/5$. A stopper is put above it and prevents undesired popping out of the cork. The temperature of the room where the champagne is opened is $T_0 = 20.0\,^\circ\mathrm{C}$, the relative humidity of the air is 70.0%, and the

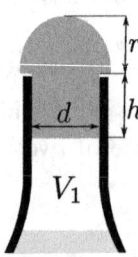

Figure 6.26 The throat of a champagne bottle

pressure is standard atmospheric. Gravitational acceleration is $g = 9.81 \frac{m}{s^2}$, while $0°C = 273.15\,K$.

(a) In the absence of a stopper, the cork would pop out when the pressure of carbon dioxide in the bottle is at least two times larger than the atmospheric pressure $p_0 = 1$ bar. Calculate the friction force between the cork and the bottle throat.

(b) A champagne bottle was cooled in a fridge so that the temperature of carbon dioxide is $T_1 = 10.0°C$, while its pressure is $p_1 = 5p_0$. Calculate the time it takes for the cork to pop out after the stopper is removed. The bottle is in the vertical position. The pressure of carbon dioxide in the bottle changes as

$$p_1(x) = p_0 \left(5 - \frac{2x}{h}\right), \qquad (6.68)$$

where x is the length of cylindrical part of the cork that is outside the bottle. Assume that the friction force is proportional to the area of contact between the bottle and the cork.

(c) What is the height reached by the cork after popping out if we assume that its motion is purely translational? Assume also that the difference of buoyancy force and air drag force is constant, that it is directed vertically downward, and that it is equal to one-tenth of the weight of the cork.

After the cork pops out from the bottle, carbon dioxide starts flowing vertically upward out of the bottle. Its speed is equal to the mean speed of carbon dioxide molecules just before popping out of the cork. The total amount of carbon dioxide flows out very fast, during $\tau = 0.800\,ms$. In a rough approximation, this process can be considered adiabatic and it can be assumed that the parameters of the carbon dioxide gas satisfy the equation of the adiabatic process at each moment during the process. The molar mass of carbon dioxide is $M_1 = 44.01 \frac{g}{mol}$, while the adiabatic index is $\gamma = 1.30$. The gas constant is $R = 8.31 \frac{J}{mol \cdot K}$.

(d) Calculate the temperature T_3 of carbon dioxide gas at the end of the process. Calculate the work A performed by the carbon dioxide gas during this process.

(e) Due to fast flowing out of carbon dioxide, the champagne bottle acquires additional apparent mass Δm. For this reason, the hands of the person holding the bottle feel a sudden push downward. The person then instinctively reacts and moves the hands upward to compensate for the push. Calculate the value of Δm.

In the course of adiabatic expansion, the molecules of carbon dioxide collide with molecules of surrounding air, including water vapor molecules. For this reason, the air near the bottle throat also cools down before thermodynamic equilibrium at room temperature is established again. Assume that the surrounding air cools down to condensation temperature $T_c = 14.5°C$ and that afterward the complete amount of water vapor in the air condenses. This is the way the small cloud that we see during champagne opening is created. The heat capacity of air is $c = 0.900 \frac{kJ}{kg \cdot K}$, its density is $\rho_v = 1.20 \frac{kg}{m^3}$, and the heat of water evaporation is $\lambda = 2.25 \frac{MJ}{kg}$. The density of water vapor in saturated air with a relative humidity of 100% at a temperature T_c is $\rho_{vp} = 17.3 \frac{g}{m^3}$.

(f) What is the volume of the cloud? Assume that the heat released during conden-
    sation of the cloud ΔQ is equal to the work A.

(g) What would be the radius of the cloud if it were spherical?

Solution of Problem 42

(a) The friction force between the cork and the bottle throat is $F_{tr} = F_{ud} - F_{vaz} - mg$, where $F_{ud} = 2p_0 S = 2p_0 \frac{\pi d^2}{4} = 41.22\,N$ is the force of carbon dioxide on bottom part of the cork, $F_{vaz} = p_0 S = p_0 \frac{\pi d^2}{4} = 20.61\,N$ is the vertical component of the force of surrounding air on top part of the cork, and $mg = \rho(\frac{\pi h d^2}{4} + \frac{2\pi r^3}{3})g = 0.05\,N$ is the weight of the cork. Consequently $F_{tr} = 20.6\,N$.

(b) The equation of motion of the cork is $m\ddot{x} = [p_1(x) - p_0]S - mg - F_{tr}(x)$. During the popping of the cork, the friction force decreases because the area of the contact between the cork and the bottle throat decreases – that is, $F_{tr}(x) = F_{tr}(1 - \frac{x}{h}) = (p_0 S - mg)(1 - \frac{x}{h})$. Using this equation and equation (6.68), we obtain the equation of motion

$$\ddot{x} + \left(\frac{p_0 S + mg}{mh}\right)x = \frac{3p_0 S}{m}, \tag{6.69}$$

which is the equation of a harmonic oscillator. Its solution can be assumed in the form $x(t) = a\cos(\omega t + \varphi) + b$. By replacing this solution in equation (6.69) we obtain the angular frequency $\omega^2 = \left(\frac{p_0 S + mg}{mh}\right)$ and the constant $b = \frac{3p_0 S}{m\omega^2} = \frac{3p_0 Sh}{p_0 S + mg}$. From initial conditions $x(0) = 0$ and $\dot{x}(0) = 0$ we obtain the amplitude $a = -b = -\frac{3p_0 Sh}{p_0 S + mg}$ and the phase $\varphi = 0$. Consequently the popping of the cork from the bottle can be described using the equation

$$x(t) = \frac{3p_0 Sh}{p_0 S + mg}\left(1 - \cos\sqrt{\frac{p_0 S + mg}{mh}}t\right). \tag{6.70}$$

Popping of the cork starts from the amplitude position $x(0) = 0$ and ends when the cork is in the position $x(t_1) = h$. The popping time of the cork is then $t_1 = \sqrt{\frac{mh}{p_0 S + mg}}\arccos\left(\frac{2p_0 S - mg}{3p_0 S}\right) = 1.71\,ms$.

(c) The speed of the cork immediately after its popping out from the bottle is

$$v_0 = |\dot{x}(t_1)| = \frac{3p_0Sh}{p_0S+mg}\sqrt{\frac{p_0S+mg}{mh}}\sin\arccos\left(\frac{2p_0S-mg}{3p_0S}\right) = 17.85\,\frac{m}{s}.$$
(6.71)

According to Newton's second law, the equation of motion of the cork in this case is $ma = F_{pot} - F_o - mg = -\frac{11}{10}mg$. Consequently after popping out from the bottle, the cork continues to move vertically upward with an initial speed of v_0 and deceleration $a = \frac{11}{10}g$, reaching a height of $H = \frac{v_0^2}{2a} = 14.8\,m$.

(d) The mass of carbon dioxide can be calculated using the ideal gas law applied to the moment when the bottle was in the fridge (state 1):

$$m_1 = \frac{p_1V_1M_1}{RT_1} = 46.8\,mg.$$
(6.72)

Immediately after the popping out of the cork, when the adiabatic expansion of carbon dioxide starts (state 2), its pressure and the volume are given as $p_2 = p_2(h) = 3p_0$ and $V_2 = V_1 + \frac{\pi}{4}d^2h$, while the temperature T_2 reads

$$T_2 = \frac{p_2V_2}{p_1V_1}T_1 = 10.20\,°C.$$
(6.73)

The pressure of carbon dioxide at the end of adiabatic expansion (state 3) is $p_3 = p_0$, while the equation of adiabatic process 2–3 gives $p_2^{1-\gamma}T_2^{\gamma} = p_0^{1-\gamma}T_3^{\gamma}$. Using equation (6.73) we find the temperature T_3 as

$$T_3 = \left(\frac{p_2}{p_0}\right)^{\frac{1-\gamma}{\gamma}} T_2 = -53.3\,°C.$$
(6.74)

The work performed by the gas during adiabatic expansion is equal to the negative change of internal energy of the gas ΔU. Since $\Delta U = mc_V\Delta T$, where $\Delta T = T_3 - T_2$ and c_V is the mass heat capacity at constant volume $c_V = \frac{R}{(\gamma-1)M_1}$, we find the work $A = -\Delta U = \frac{m_1R(T_2-T_3)}{(\gamma-1)M_1} = 1.87\,J$.

(e) The apparent increase of the mass of the bottle comes from the reactive force F_r that acts on the bottle due to flowing out of the champagne. This force is analogous to the force that moves a rocket as a consequence of flowing out of the fuel. Since the complete amount of carbon dioxide of mass m_1 flows out during time interval τ at a speed v, it follows that $F_r = \frac{m_1v}{\tau}$. It was stated in the problem that the speed of the gas is equal to the mean speed of the molecules before popping out of the cork $v = \sqrt{\frac{8RT_2}{\pi M_1}}$. Consequently the apparent mass is equal to $\Delta m = \frac{m_1v}{g\tau} = 2.20\,kg$.

(f) During collisions of carbon dioxide molecules and the molecules of surrounding air, the air cools from temperature T_0 to temperature T_c. Next the water vapor condenses and the drops of water that form the cloud appear. Consequently

$$\Delta Q = m_2\, c\, (T_0 - T_c) + \lambda m_{vp}, \qquad (6.75)$$

where $m_2 = \rho_v V$ is the mass of air and water vapor in volume V, while m_{vp} is the mass of water vapor in the same volume. The density of water vapor with relative humidity of 70% is equal to 70% of the density of water vapor of relative humidity 100% at the same temperature, which leads to $m_{vp} = 0.7\rho_{vp}V$. We assume, as stated in the text of the problem, that $\Delta Q = A$, which leads to $V = \frac{A}{\rho_v c (T_0 - T_c) + 0.7\lambda\rho_{vp}} = 56.3\,\mathrm{cm}^3$.

(g) If the cloud were spherical, its volume would be $V = \frac{4}{3} r_1^3 \pi$. Consequently its radius is $r_1 = \sqrt[3]{\frac{3V}{4\pi}} = 2.38\,\mathrm{cm}$.

We refer interested readers to reference [37].

References

[1] R. M. Alexander. Simple models of human movement. *Appl. Mech. Rev.*, 48:461–470, 1995.

[2] M. H. P. Ambaum. *Thermal Physics of the Atmosphere*. John Wiley and Sons, 2010.

[3] L. A. Bloomfield. *How Things Work*. John Wiley and Sons, 2010.

[4] P. J. Brancazio. Physics of basketball. *Am. J. Phys.*, 49:356–365, 1981.

[5] J. W. M. Bush and D. L. Hu. Walking on water: Biolocomotion at the interface. *Annu. Rev. Fluid Mech.*, 38:339–369, 2006.

[6] I. Cantat, S. Cohen-Addad, F. Elias, et al. *Foams, Structure and Dynamics*. Oxford University Press, 2013.

[7] R. Cross. The dead spot of a tennis racket. *Am. J. Phys.*, 65:754–764, 1997.

[8] R. Cross. Effects of friction between the ball and strings in tennis. *Sports Eng.*, 3: 85–97, 2000.

[9] P. Davidovits. *Physics in Biology and Medicine*. Elsevier, 2008.

[10] C. Frohlich. Effect of wind and altitude on record performance in foot races, pole vault, and long jump. *Am. J. Phys.*, 53:726–730, 1984.

[11] T. Garrison. *Essentials of Oceanography*. Brooks/Cole, 2009.

[12] A. Gonzalez, J. B. Gomez, and A. F. Pacheco. The occupation of a box as a toy model for the seismic cycle of a fault. *Am. J. Phys.*, 73:946–952, 2005.

[13] L. Gunther. *The Physics of Music and Color*. Springer, 2012.

[14] J. Güémez and M. Fiolhais. Forces on wheels and fuel consumption in cars. *Eur. J. Phys.*, 34:1005–1013, 2013.

[15] O. Helene and M. T. Yamashita. A unified model for the long and high jump. *Am. J. Phys.*, 73:906–908, 2005.

[16] I. P. Herman. *Physics of the Human Body*. Springer, 2008.

[17] H. Hunt. Giroscopes and boomerangs. www3.eng.cam.ac.uk/~hemh1/boomerangs .htm.

[18] H. Kanamori and E. E. Brodsky. The physics of earthquakes. *Phys. Today*, 54:34–40, 2001.

[19] S. Kinoshita, S. Yoshioka, and J. Miyazaki. Physics of structural colors. *Rep. Prog. Phys.*, 71:076401, 2008.

[20] H. Lin. Fundamentals of zoological scaling. *Am. J. Phys.*, 50:72–81, 1982.

[21] J. O. Linton. The physics of flight: I. fixed and rotating wings. *Phys. Educ.*, 42:351–357, 2007.

[22] G. Margaritondo. Explaining the physics of tsunamis to undergraduate and non-physics students. *Eur. J. Phys.*, 26:401–407, 2005.

[23] W. A. Michael. Using physics to investigate blood flow in arteries: A case study for premed students. *Am. J. Phys.*, 78:970–974, 2010.

[24] H. M. Nussenzveig. Theory of the rainbow. *Sci. Am.*, 237:116–127, 1977.

[25] P. Goldreich, S. Mahajan and S. Phinney. Order-of-Magnitude Physics: Understanding the World with Dimensional Analysis, Educated Guesswork, and White Lies. Draft of 1999. www.inference.org.uk/sanjoy/oom/book-letter.pdf.

[26] E. G. Pita. *Air Conditioning Principles and Systems*. Prentice Hall, 2002.

[27] S. A. Ramakrishna and T. M. Grzegorczyk. *Physics and Applications of Negative Refractive Index Materials*. Taylor and Francis, 2009.

[28] L. Salkin, A. Schmit, P. Panizza, and L. Courbin. Generating soap bubbles by blowing on soap films. *Phys. Rev. Lett.*, 116:077801, 2016.

[29] C. H. Scholz. *The Mechanics of Earthquakes and Faulting*. Cambridge University Press, 1990.

[30] R. Shepard. *Amateur Physics for the Amateur Pool Player, 3rd edition*. 1997. www.sfbilliards.com/Misc/Shepard_apapp.pdf.

[31] W. Shockley and H. J. Queisser. Detailed balance limit of efficiency of p-n junction solar cells. *J. Appl. Phys.*, 32:510–519, 1961.

[32] B. M. Smirnov. *Fundamentals of Ionized Gases*. WILEY-VCH, 2012.

[33] V. M. Spathopoulos. A simple model for ski jump flight mechanics used as a tool for teaching aircraft gliding flight. *e-Journal of Science and Technology*, 5:33–39, 2010.

[34] V. M. Spathopoulos. A physics heptathlon: Simple models of seven sporting events. *Phys. Educ.*, 45:594–601, 2010.

[35] G. P. Sutton and O. Biblarz. *Rocket Propulsion Elements*. John Wiley and Sons, 2001.

[36] M. Vollmer. Physics of the microwave oven. *Phys. Educ.*, 39:74–81, 2004.

[37] M. Vollmer and K.-P. Mollmann. Vapour pressure and adiabatic cooling from champagne: Slow-motion visualization of gas thermodynamics. *Phys. Educ.*, 47:608–615, 2012.

[38] J. Wesson. *The Science of Soccer*. IOP Publishing, 2002.

[39] S. Yoshioka, E. Nakamura, and S. Kinoshita. Origin of two-color iridescence in rock dove's feather. *J. Phys. Soc. Jpn.*, 76:013801, 2007.

[40] H. D. Young, R. A. Freedman, T. R. Sandin and A. Lewis Ford. *Sears and Zemansky's University Physics with Modern Physics*. Addison Wesley, 2003.

[41] M. Zamir. *The Physics of Coronary Blood Flow*. Springer, 2005.